THE LAST TRUE PRESIDENT

By Benjamin Casad

Table of Contents

INTRODUCTION

The current situation, both in America and worldwide, reminds me of a scene in the movie *The China Syndrome*. The Jack Lemmon character had been warning about the defective, unsafe engineering that the corrupt corporate leaders had imposed on the California nuclear power plant. The corporate leaders had forced a test of the plant's safety system that Jack Lemmon believed was not only unsafe but could lead to a catastrophic, uncontrolled reaction. As the test progresses, alarms and loud sirens go off, and the control room goes dark except for the flashing red lights. The plant appears to be on the verge of a catastrophic meltdown, and many there seem to realize, albeit too late, that perhaps Jack Lemmon was right after all. Perhaps it wasn't a good idea to risk a catastrophic meltdown.

Our current national situation reminds me of that scene. We already see the red lights flashing, alarms sounding, and the room going dark. The list of warning signs goes on and on: A new trillion-dollar debt added every three months; tens of thousands of hostile military forces, terrorists, and criminals currently present on American soil; drug cartels earning billions in American cash and gaining substantial control of our border; the surrender of billions of dollars in arms and military equipment to terrorists and hostile militaries; tens of millions of illegal immigrants brought into the country; thousands of youths dying from fentanyl and other drug overdoses; armed criminal gangs taking over apartment buildings in several cities without objection from public authorities; Chinese military and police presence on American soil and surveillance balloons violating our airspace. Finally, in July of 2024, there was an attempted assassination of President Trump that may have involved cooperation from elements within the deep state. Why are we, at the national leadership level, accepting radical and unnecessary risks? Why are we risking the peace and security of the American Republic? Instead of building back better, why not leave the American Republic intact? We should be very cautious about willingly destroying or even risking our priceless and irreplaceable inheritance—our Constitutional Republic.

The conclusion reached in this book may be disturbing to some—that the United States suffered a *coup d'état* in 2020. America suffered a *coup d'état* that was perpetrated most obviously by the fraudulent national elections in 2020. We began to cross the Rubicon, though very few Americans of either party wanted that or were even aware of what was happening. The election of 2020 and its aftermath created a Constitutional crisis that is still unresolved. The 2020 and 2024 elections should be interpreted as the potential end of our Constitutional era, not merely another election.

The 2024 election may be the last national election decided by American citizens. Indeed, maybe 2024 won't even be decided by American citizens. The Democrats are actively and aggressively seeking to extend voting rights to non-citizens and illegal immigrants, even for the 2024 election. This plan has been obvious for some time. The basic plan is to effectively replace citizen sovereignty with non-citizen sovereignty without passing laws that specifically grant the right to vote to non-citizens. The Democrats know that the only way they can possibly "win" is to surrender the sovereignty of American citizens by eliminating citizenship as a meaningful legal status.

If President Trump is not restored to the Presidency after the 2024 election, there will never be another meaningful, legitimate presidential election. There will never be another true President. America will enter a post Constitutional era. There will continue to be an imitation of Constitutional processes in the

United States and a parody of treating the Constitution as our ultimate source of law, but the true meaning and authority of the Constitution will have been eviscerated. The real power will have effectively been removed from the hands of the citizens.

America will suffer a catastrophic blow from which recovery may not be possible. The stakes in the 2024 election are not simply the future of the American Republic. At stake is the continued existence of the American Republic. This is what the President means when he says we won't have a country anymore. He is absolutely right.

As 2024 dawns, the nascent totalitarian nature of the Biden-Harris/deep state regime is obvious. Biden's handlers are doing everything they can to eliminate free and fair elections both within the Democratic Party and in the broader general election. They have actively rigged the Democratic primary to prevent Robert Kennedy from gaining the nomination. They forced Biden out and installed Kamala Harris without any democratic process. Why wouldn't they do the same thing to the rest of us to prevent us from re-electing President Trump? Of course, they would, and they are.

The massive lawfare they have initiated against President Trump, and especially the efforts to remove the President from the ballot in numerous battleground states, are just a few examples of the direct attacks on the democratic process and the American Republic. Thank God they have so far failed, but the more significant point is that they have tried. These efforts are part of a conscious, willful design to prevent the American people from choosing their President and re-establishing control over the American government. When Joe Biden persecutes his main political opponent through lawfare, both civilly and criminally, what further proof is needed for reasonable minds to conclude that the rule of law in the United States is on life support? What further proof is needed for reasonable minds to conclude that Biden is acting in a manner that is totally inconsistent with a good faith effort to execute his duty to preserve, protect, and defend the Constitution and sovereignty of the United States?

We have seen this same type of operation in other contexts before. The leftist and progressive globalists are using authority derived from democratic and quasi-democratic processes as a foundation to eliminate the opposition and erode the efficacy of popular sovereignty and the Constitution. The Marxist elements have already announced their plan for lawfare following the 2024 election. If they somehow gain control of the House, they will provoke a Constitutional crisis by obstructing the certification of President Trump's election. Can there be any more obvious statement of their true intentions than that? These leftist elements have no intention of acting in good faith to promote the orderly functioning of our Constitutional Republic. They would burn the Constitution just as quickly as they burn the American flag.

The American Republic itself is on life support. It is not hyperbole or exaggeration when the President says 2024 is likely our last opportunity to save the Republic from collapse and totalitarianism. And no, the threat is not from so-called MAGA extremists who have been falsely labeled as white supremacists and insurrectionists. The threat is from leftists and progressive globalists—all those who believe America should be "transformed" from a free sovereign Republic based on the sovereignty and consent of the citizen electorate, and "build back better" through a process they call progressive social reconstruction. Many of the leftists and globalists are well-intentioned people, unaware of the true danger, and are being used by progressive globalist leaders and their massive dragnet of media, academia, big tech, a woke and weaponized bureaucracy, virtually unlimited funding, and woke organizations and bureaucracies within the U.N., E.U., and other international bodies.

Throughout the world, and certainly in every Western country, people are confronted with efforts to undermine national sovereignty and individual human and property rights. These efforts include mass weaponized migration; international controls on matters that were formerly decided by sovereign states, such as migration, pollution and carbon emission and controls related to community health and other matters. In the case of the United States, immigration was weaponized by Biden in violation of domestic law.

Globalist doctrine openly espouses and welcomes a post-human world. Thanks, but no thanks. A new world order will inevitably be dominated by criminal or corrupt elements, spiritually bankrupt people or those who are not motivated by a sincere, good faith commitment to protect the true core principles of a human rights-based civilization. These core principles include; basic fundamental human rights; religious freedom; social policies that protect human and family rights; property rights; the rule of law and legal processes consistent with God-given human rights and human dignity.

The true cost of the loss of our Republic will be our eventual surrender to a new world order. The new world order will probably, eventually, be under the control of the Chinese Communist Party and other corrupt and criminal elements that do not share pro-human values.

There is a nexus, a common thread, that ties together political, economic, and human rights. That thread is the rule of laws promulgated by free civilization—civilizations that rule by the legitimate consent of the governed. The vast majority of people throughout the world want peace, freedom, stability, prosperity and social policies that foster strong families and human rights.

The existence and strength of a human rights-based international order ultimately depends on the stability and strength of sovereign nation states. Strong sovereign nation states produced by free civilizations that support and value human rights and human dignity are essential to resisting corruption and preserving a human rights-based international order. Ultimately, national sovereignty and human rights are inseparable. Human rights worldwide cannot be secure if the United States is destabilized from within.

One of the biggest problems with our U.S. economy is that we continue to refuse to live within our means. We have allowed powerful entrenched interests, including some in the Republican Party, to misappropriate the wealth of the American people and endanger the viability of our currency. By continuing to print money and monetize our debt under the fiction that we are "borrowing" money from the Federal Reserve and private holders of U.S. bonds, we avoid paying the costs in the near term. The true cost is paid by the devaluation of our currency and the theft of wealth, opportunity, and consumer welfare from present and future generations, who will be saddled with unpayable debt. The end result of these policies will be the destruction or severe devaluation of our currency. This ongoing exploitation of our currency has created enormous political pressure on the fragile ability of free civilizations to maintain self-rule and pro-human policies.

There are powerful elements in both parties in the U.S. and internationally who want unfettered access to the full faith and credit of the United States—also known as American citizens, their property, and their descendants' property—to finance whatever they want. This is why progressive globalists vehemently object to any efforts by American citizens to regain control of the U.S. dollar. This is the main reason the deep state has announced, through Joe Biden, that Donald Trump must never return to the Presidency.

The destruction of our currency is, in effect, a taking of property from all of us who hold dollars. One slight problem with destroying the dollar—it is not just a taking of property; it will almost certainly lead to severe stress or the collapse of our civil society.

The policies of Biden and his deep state enablers have created an almost perfect storm of catastrophes, virtually all of them self-created. Biden and Harris have willfully allowed tens of thousands of terrorists and hostile military elements to enter the U.S. There is a crisis of failing energy production, massive crime combined with a self-inflicted reduction of law enforcement and prosecution, reductions in food supplies, radical inflation and exploitation of the dollar, a totally unnecessary flirtation with World War III, and ultimately an unnecessary risk of the collapse of our Constitutional Republic.

There is a significant issue as to whether the actions of Joe Biden and his enablers are intentional or merely the result of gross incompetence and corruption. This book takes the position that their actions are willful and intentional and not the result of good faith incompetence. The problem goes far beyond Joe Biden. He personally may well imagine that he is doing good for the American people. It is not due to a failure of the plan that our sovereign Constitutional Republic is on the verge of catastrophe or even collapse. "Transformation" is the plan. This issue is significant because the conclusion the citizen electorate reaches regarding the true motives of Joe Biden and his enablers will guide the electorate in empowering political leaders to craft an appropriate response.

Joe Biden and his enablers' true intentions can be discerned from their actions in several key areas. These areas include: the weaponization of immigration; the intentional weaponization and subversion of our legal and electoral system; the protection of election fraud and violations of election finance laws; the institutionalized theft of wealth and the destruction of our currency; the willful surrender of American sovereignty in numerous areas; the surrender of billions of dollars in arms and military equipment to terrorist organizations; and knowingly allowing criminal drug and human trafficking cartels to reap billions of dollars in illegal income. None of these developments happened because of a good faith exercise of political power. They are all part of an effort to transform the American Republic under the color of law and to ultimately eliminate citizen sovereignty permanently from the American political process.

If the American Republic is to survive, which is by no means a foregone conclusion, it will be in part due to independents, true progressives, young people, African Americans, other minorities, moderate Republicans, and especially the millions of Americans who identify as Christian but don't vote, recognizing that our Republic is facing an existential threat emanating from both internal and external forces.

There are many of us who are natural allies in the cause to preserve our civil society and self-rule. These groups include: those in the pro-human camp—team humanity; the liberal progressives and others who truly don't want to destroy the American Republic, such as Robert F. Kennedy Jr. and his supporters; those who believe in fiscal sanity; those who believe America is somehow racist but don't believe complete self-destruction is wise; the never-Trump elements within the Republican Party who imagine that populism and nationalism are somehow inconsistent with conservative values and principles, or those who have decided they aren't concerned enough about the potential collapse of the American Republic to actually support President Trump. Almost anyone who is not dedicated to the destruction of the American Republic should be honest enough to recognize the seriousness of the crisis we are in and act accordingly. If we recognize the danger we are all in, we can and should work together to prevent it from happening. The self-inflicted collapse of America's currency, civil society, and rule of law is not in the interest of the American people. It is also not in the interest of world peace or the people of the world. It is not in the interest of minorities, trans kids, gay people, or any of the so-called oppressed groups.

This book is an attempt to reach out in a spirit of peace and reconciliation to sincere pro-human, peace, and freedom-loving people in the U.S. and the world over to consider what is truly at stake. This book is an alarm, a wake-up call in the middle of the night to urge every one of good faith who wants free civilization and meaningful rule of law to survive both in the U.S. and worldwide to recognize the extreme danger that we are all in and to put aside our differences, which are petty in comparison to the urgent need to secure a pro-human agenda both domestically and worldwide.

This book is also an account for future generations of Americans and freedom-loving people throughout the world so that others can have a perspective to appreciate the seriousness and complexity of the crisis we, the American people, faced in both the dark years of 2020 to 2024 and in the monumental election of 2024. This book is an account so that others can know that the American people fought back and refused to surrender their Republic without a fight.

Finally, thank you, President Trump, to you and your family, for standing up and fighting for the American people in this very dangerous moment, both for the world and the American Republic. Your persecution by a party and regime that no longer has anything of merit or value to offer the American people is proof of their unworthiness and the existential threat they truly pose to our Republic. The courage and perseverance you have shown in bearing the brunt of this attack on our Republic is proof that you are the one who deserves the sacred trust of the American people.

CHAPTER ONE

2020

The years 2019 to 2020 were a time of monumental events worldwide and certainly were catastrophic in the United States. In early 2019, President Trump was heading toward almost certain re-election. The economy was doing well and there was a plan in place to wind down the war in Afghanistan. Late in 2019, news reports began to surface of a mysterious respiratory disease that was causing fatalities. The disease caused pneumonia-like symptoms. People with weak or compromised immune systems were susceptible to being overcome by COVID-19 if their immune systems were not strong enough to kill the infection. The media and many governments worldwide aggressively perpetrated the myth that COVID-19 had natural origins. It took until late 2023 or even 2024 for the media to quietly and effectively admit that COVID-19 was manmade.

By January of 2020, there were mandatory orders to lock down commerce and the movement of people worldwide under the fiction that such actions would help stop the spread of COVID. The Democrat Party in the United States began a massive lawfare campaign designed to expand mail-in balloting and also reduce ballot security measures such as voter ID, signature verification, accurate voter registration rolls, drop boxes, and same-day registration to make it more difficult to cull illegal votes.

We have 2 main political parties in the United States. In 2020, America faced a profound decision. For many years or even decades, we have been able to avoid the hard choices. The election of Donald Trump in 2016, the last true President, began to put into very clear focus the stark contrast that truly exists between the 2 parties on a wide range of issues. The 2 parties' goals are very different and frankly, incompatible and nearly irreconcilable with each other. As this is written, we don't know how this clash of titans will turn out.

The division was only broadly speaking between Democrats and Republicans. More accurately, the division is between globalists, of which there were many in the Republican party, and nationalists represented by Donald Trump. The time-honored American principles of compromise and reaching out to the other side were becoming increasingly difficult. More and more American people were beginning to conclude, "Why do I want to build bridges to a place I can't go." We cannot compromise with people who are acting in bad faith and who actively want to destroy our Constitutional Republic.

This current Democrat Party is simply not the Democrat Party of old. There was a day when Democrat Party leaders were patriots and acted in good faith. People like John and Robert Kennedy, Hubert Humphrey, and even Jimmy Carter. Those days are long gone and they will probably never be coming back. The Democrat Party has been gradually taken over by progressive globalists and other left-wing elements. This Democrat Party no longer represents the American people in a meaningful sense. The announcement of Robert F. Kennedy Jr. in August of 2024 endorsing President Trump made dramatically clear the urgent need to prevent the forces that are using and controlling the Democrat Party from continuing to hold political authority. The current Democrat Party is one of the vehicles that progressive globalists have used to bring the American Republic to the brink of self-destruction.

The current Democrat Party has, for all intents and purposes, merged with the deep state. The Democrat Party is the political arm of the deep state. There is a symbiotic relationship between the current Democrat Party and the mainstream media, big tech, the deep state, the woke and weaponized bureaucracies, and radical Non-Government Agencies. These groups have different roles and tasks but their basic goals are very similar— to facilitate progressive social reconstruction and the "transformation" of the United States; to eliminate citizen sovereignty; to remove and neutralize the United States as an effective barrier to the advent of a new world order and to suppress or censor unapproved information and any political, social or cultural opposition.

The current Democrat Party also represents the elite globalist international elements from Davos and Brussels and the uber-elite progressive billionaires that have provided virtually unlimited funding to achieve shared goals.

The term globalist is used here not to describe a small cabal of conspirators but to describe parties and elements throughout America and the world who had a convergence of interests, especially in the 2020 U.S. election. To the domestic "progressives", the Marxist elements, Antifa, BLM, and the leadership of the Democrat Party under Barak Obama the immediate goal was the continued "transformation" of the

American Republic from a free sovereign Republic into a 1 party Marxist oriented dictatorship along the lines of Venezuela, Cuba or Nicaragua.

Their program was to facilitate a "transformation" under the color of law using free institutions such as elections, media, and information, weaponized immigration, and the weaponization of our justice system and bureaucracy to achieve their goals. The program of transformation also included attacks on regime opponents and an independent judiciary.

The progressive/globalist plan motivated by numerous slogans such as; "build back better," "managed decline," "equity," "we won't go back," and now "turn the page" is a plan to purposefully destabilize the United States and the free world. The progressive/globalist agenda is quite out in the open through the publications of the WEF, World Economic Forum, and myriad U.N. agencies and NGOs.

The foundation of this agenda is the elimination of national sovereignty in the Western world through weaponized migration. Our allies in Europe, the United Kingdom, Australia, New Zealand, and Canada have all been damaged to some extent by adopting progressive/globalist policies, especially weaponized migration.

The end result of these policies will be to eliminate meaningful national sovereignty and citizenship in Western Europe, the U.K., Australia, Canada and the United States by flooding western countries with tens of millions of illegal immigrants, terrorists, and hostile military elements.

There are many other policy programs that are part of the same plan for progressive social reconstruction. These programs include; degrading and dismantling manufacturing and energy supplies in the Western world, particularly in the United States; degrading the energy infrastructure and food supplies; devaluing and exploiting the dollar to the point of collapse; executing a large-scale purge of patriotic and pro-American elements from the military and gradually surrendering American sovereignty to U.N. bureaucracies, such as the World Health Organization (WHO), with the ultimate goal of eliminating the sovereignty of the American Republic.

This destructive plan to "transform" our Constitutional Republic is totally inconsistent with fundamental principles of ordered liberty, good faith, citizen sovereignty, and a meaningful social compact and is also inconsistent with the security and survival of the United States of America as a sovereign Constitutional Republic.

The progressive globalist plan, either by design or by default, would embrace the defeat of American sovereignty as not only inevitable but also a positive good. They tell us that there should be a managed decline in America. America should "chart a new course" for a post-Constitutional era. The geographic area that formerly was the United States of America can then be integrated into a new world order ostensibly under the aegis of the United Nations. Peace and world order can be achieved through the transformation of American sovereignty.

There is a fundamental principle of progressive globalism that a safe and orderly emergence of a new world order is possible within the context of the managed decline of the American Republic. According to this premise, a new world order will emerge that will be a rules-based international order that respects human rights and human dignity and will also be independent of the Chinese communist party and secure from any military threat. Voluntary integration of the United States into a new world Order will enable the neo Marxist elites in the United States to have some influence in the emerging new world Order. How exactly that would happen is not completely clear, but we will surely build back better following the inevitable collapse of the dollar and the civil society in the American Republic.

Once we accept as a matter of public policy that anyone in the world has a right to move to the United States, be supported at crippling public expense and be granted voting and electoral rights and political authority, we have already surrendered our sovereignty both as citizens and as a nation. We then lose control over what happens to our county. How will we ever regain control of our country? If there could possibly be anything more reckless and irresponsible than allowing soldiers and terrorists from hostile countries to enter the United States it would be to allow them to vote and receive public support while they are awaiting activation. This is what Joe Biden, Kamala Harris, and their controllers are attempting.

You may ask yourself why they are doing this. Why is this plan consistent with a good faith stewardship of the American Republic? Well, the answer is it isn't. They are not acting in good faith. They have no intention of securing and ensuring the American Republic's survival. Their intention is to "transform" the American Republic to "turn the page" and to implement a plan of progressive social reconstruction. This is not to suggest that rank and file Democrats actively want to destroy the American Republic. Most of course do not and many Democrat Party office holders may very well imagine that they are acting in good faith on behalf of the American people. The difficulty is that the polices pursued by the national party leadership are clearly inconsistent with the sovereignty and survival of the American Republic.

One thing globalists and nationalists appear to agree on is that the globalist agenda is completely incompatible with a sovereign, strong, free, solvent, and stable American Republic. There is no way to implement a one world government or even mass immigration and binding United Nations mandates on a wide range of issues as long as there is a resolute American Republic in the world that is willing and able to resist globalist coercion and U.N. mandates.

The Nationalist vision, as articulated by Donald Trump, is very different. Donald Trump openly and boldly rejected almost the entire progressive globalist agenda. Donald Trump rejected the very notion that it is the United States government's duty to create 'equality' or even to fight climate change.

To Donald Trump, managed decline is not a recipe for peace and security. The transformation of America is a recipe for collapse that will make world war a virtual certainty as the world fights over the carcass of the American Republic and, ultimately, the geographic area that was once the United States of America. President Trump and his nationalist agenda rejected the very notion that American sovereignty can safely be surrendered either to the U.N. to domestic enemies of the Republic or a new world order.

To Donald Trump, a new world order is not the solution—it is the problem that inevitably will lead to the collapse of America as we have known it and the advent of a globalist tyranny.

Trump believed in fiscal sanity, a strong dollar, and the American people asserting control over their currency. Trump wanted fair trade deals and supported tariffs and reciprocal policies to achieve fair trade deals. He wanted real, actual control over the borders. He believed in a strong United States military. Although Trump was not opposed to alternative energy sources, he did not want sovereignty and money to be surrendered to international bureaucrats in the effort against climate change.

Donald Trump insisted that the financial exploitation of America and the American dollar must stop. The exploitation of the American dollar was caused primarily by internal policies—deficit spending, the insatiable consumption of money by the deep state, and the steadfast, adamant refusal of American elites to live within our means. Trump believed that our massive trade deficits were partly caused by unfair and outdated trade deals which caused unnecessary strain on the dollar's value by expanding our trade deficits.

American nationalism is not isolationist or opposed to either a rules based international order or to collective security measures. America had made herculean efforts and taken huge risks to support a rules based international order and would continue to do so. Trump believed that America had to act quickly and urgently to protect its markets and borders, rebuild its energy supply and military, institute fiscal sanity to strengthen its currency and insist on fair and equitable trade and security agreements. Through following these policies, America could secure itself, deter potential aggression and war, and be a more effective bulwark for the free world against globalist totalitarianism.

Donald Trump was not afraid to say that the policies advocated by progressive globalists were not good faith policies propounded by people who sincerely have the interests of the American Republic at heart. To Donald Trump, there is no way an American President acting in good faith could fail or refuse to work towards the goals or securing the peace, sovereignty, security and stability of the American Republic.

Trump articulated the feeling that America can no longer afford the luxury of complacency and of granting political authority to people who either do not want or do not care if the American Republic survives. Regardless of their true motives, whether due to gross incompetence or willfulness, those who are indifferent to the survival of the American Republic must not be allowed to hold political authority.

Some of the things Donald Trump said seemed extreme when compared to the mainstream media narrative that the American people have been told over and over again. The media and big tech have succeeded in convincing many American people and others throughout the world that Trump and his rejection of progressive/globalist policies is dangerous. The absurd and slanderous lie that Trump is racist has become deeply embedded in the minds of American people who take the mainstream media seriously. Many intelligent, well-educated, liberal-minded people have closed their minds to the possibility that the media and big tech may be lying to us about Trump and a whole range of other issues.

In May of 2024 the United States, through the Biden deep state, was pushing a proposal of acceding to a WHO (World Health Organization) resolution which would function within the U.N.

framework as a treaty even though the U.S. Senate would never approve it. This resolution would significantly impair the national sovereignty of not only the United States but of other nations as well. The resolution would enable U.N. bureaucrats to declare an "international health emergency" and be able to force sovereign countries to adopt measures like vaccine mandates, digital IDs, gun confiscation, passport controls and even possibly controlled digital currency. The U.N. and other international bureaucracies have already encroached on the sovereignty of the United States and many other countries. National sovereignty has been or could be undermined through controls on public health regulation; climate change agenda controls; health-related passport controls; digital IDs and financial controls through controlled digital currency. National sovereignty is currently eroding throughout the Western Judeo-Christian world due to weaponized migration.

President Trump withdrew the United States from the WHO for a very good reason. The WHO is a globalist bureaucracy significantly controlled by the Chinese communist party and significantly paid for by the United States. This withdrawal did not mean the United States was unconcerned about international health issues. The withdrawal meant that the United States would not surrender its sovereignty regarding health issues to an international bureaucracy.

The Presidential debate in late June of 2024 revealed the mainstream media's willingness and capacity to lie to the American people about even the most obvious truths. In the debate, the American people and the world were suddenly confronted with an appalling situation. We realized we have a patient suffering from serious cognitive issues sitting in the White House and posing as the President of the United States. How could this have happened? Why weren't the American people told of this? If the media has been lying to us about something as obvious as Biden's capacity, what else are they lying about? Could they have lied about COVID? Could they have lied about the 2020 and 2022 elections? Could they be lying to us about Donald Trump? Could they be lying about January 6th, 2021? Could they be lying about the invasion of our country? Could they be lying about the Ukraine?

When we deconstruct the mainstream media narrative and compare it to real facts, we can see the mainstream media narrative is an orchestrated cacophony of propaganda, misinformation, or lies ultimately based on propaganda and is not a credible basis for policy decisions. The mainstream media narrative, therefore, is not a reliable source of information on which the citizen electorate can decide who should be granted political authority.

CHAPTER TWO

WEAPONIZATION OF IMMIGRATION

The cornerstone of progressive social reconstruction and the foundation of the plan to transform America is the effective elimination of citizen sovereignty. One of the key tools used to accomplish America's transformation is massive illegal immigration or weaponized migration. Organized massive illegal immigration will ultimately vest decisive political authority in non-citizens and illegal immigrants and enable the progressive globalists to establish a 1 party dictatorship within the United States. Non-citizen political and electoral support could easily in the near future, if not now, be enough to tip the scales in favor of the Democrat party.

Progressive social reconstruction will ultimately have to eliminate Citizen sovereignty because Citizens might not consent to America's transformation. Citizens cannot be relied on to stay the course towards progressive social reconstruction. Citizens might not consent to some of the collateral consequences of America's transformation, such as the loss of our sovereignty, borders, and markets, the institutionalized theft of wealth and the loss of our currency, and ultimately, the loss of control of our military and our country. Citizen non-consent could ultimately sidetrack or even defeat the plan of transformation altogether. Citizens might ultimately refuse to finance progressive social reconstruction.

The elimination of citizen sovereignty is vitally important to the goal of progressive social reconstruction because it will effectively ensure perpetual one-party rule. One-party rule will mean the deep state will never be accountable to the tax-paying electorate. The citizen electorate will never be able to remove or reform the deep state. Despite the massive and obvious security risks of allowing terrorists, criminals, and hostile military forces to infiltrate the United States, the Democrat Party leadership has concluded that as long as 1 party rule can be secured, it doesn't matter what the eventual cost to the American people is. American citizenship will be unnecessary in a post-constitutional America because we will all be global citizens. American citizenship will not only be unnecessary it will ultimately be subversive.

There is no way progressive globalists can guarantee their political power and their ability to control America's wealth if they have to survive free, fair, and auditable elections among a citizen electorate. America's social reconstruction cannot be completed, and we cannot "turn the page" if the electorate is free to withhold its consent. Or, worse yet, free to actually refuse to finance progressive social reconstruction.

Biden and Harris moved immediately to open the borders and facilitate a literal invasion of the country. The main purpose of facilitating an invasion of the country is to transfer political power from American citizens and to vest decisive political power with non-citizens. When they talk about the transformation of America, this is what they mean—removing control of the American Republic from the hands of American citizens. President Trump had warned the American people that Biden and his enablers would do exactly that if they could. The media had, of course, refused to disclose this plan to the American people before the 2020 election but it was hardly a secret to informed people.

There are several other purposes for the weaponization of immigration and for flooding the country with illegal immigrants. First, it will distort the population base of citizens necessary to support a Congressional seat or electoral vote. Apportionments are currently made on the basis of population, not citizens. Even if illegal immigrants are not immediately allowed to vote, they will affect apportionment and electoral votes. And, many illegal immigrants will eventually vote.

Citizenship is very important and inexorably related to sovereignty. In our current situation, we are granting electoral votes to vast numbers of illegal immigrants whose only stake in America is the benefits they receive. Many of the immigrants are good people who may make real contributions to our society. No one denies that. The issue is the process of how immigrants become lawful residents and citizens.

Massive illegal immigration is another example of an old Democrat Party play—using public resources and money to fund and support its political base. Illegal immigrants will consume massive amounts of public resources. These resources include housing, health care, education, welfare, and of course, social security. Although currently nearly bankrupt, our social security system could be salvaged if no significant new burdens are placed on it. Many illegal immigrants and non-citizens will be supported at massive public expense and become substantially dependent on big government and the deep state. Weaponized illegal immigration is a public investment in the present and future political power of the Democrat party. This investment will be paid for by what remains of the dollar and by citizens who will ultimately suffer a reduction or loss of their sovereignty.

Non-citizen elements of the body politic will, by and large, support the Democrat Party and allow the Democrat Party to establish a one-party totalitarian state in what will effectively become a post-Constitutional America. The deep state and the bureaucracy would become permanently entrenched and would be able to misappropriate the wealth of the American people on a macroeconomic scale even more than they already do.

International globalism as represented by the U.N. also encourages massive immigration. In this respect, U.N. goals align with those of the Democrat party. The U.N. has been providing funding for millions of illegal immigrants who cross both the U.S. and other borders. The United States provides about 25% of the funding for the UN. The United States also of course provides massive funding to house and support tens of millions of illegal immigrants.

There are many other tools that are currently being used in the process of eliminating citizen sovereignty and the meaningful consent of the governed. These tools include; the vigorous suppression of free speech through government intervention and manipulation of media and big tech; the institutionalized theft of wealth; massive wealth redistribution and ceaseless lawfare initiated in bad faith on many fronts.

The Democrat Party also wants to grant the right to vote to tens of millions of non-citizens and illegal immigrants. Even if they do not succeed in getting the legal right to vote for illegal immigrants in 2024, many illegal immigrants and non-citizens will be allowed to register to vote and still send in mail-in ballots anyway and clog the system with illegal ballots that may result in another unverifiable and uncertifiable "election". Approximately 9 to 14 % or more of non-citizens and illegal immigrants voted in the 2020 U.S. elections. This number will almost certainly go up even if reasonable election security measures are enacted.

The issue of whether non-citizens vote in significant numbers is somewhat controversial. Some say non-citizen voting is virtually nonexistent. Others claim it is a real issue and increasingly more difficult to regulate and significant numbers of illegal immigrants and non-citizens register and vote. Many non-citizens did vote in the 2020 U.S. elections. Virginia recently removed over 6,000 non-citizens from its voting rolls. In Texas, the attorney General attempted to remove a significant number of non-citizens from the voting rolls. In Texas, there have been very aggressive efforts to preserve non-citizens on the voting rolls and to prevent the Texas Attorney General from removing non-citizens from the voting rolls. One thing is for sure: Any effort to prohibit or outlaw the registration of non-citizens or to cull them from voting rolls for Federal elections is vigorously opposed by Democrats. The lawfare is initiated by both the Department of Justice in Democrat administrations and private lawfare initiatives.

The Democrat and deep state plan to effectively grant the franchise to tens of millions of illegal immigrants and non-citizens is in direct contravention of both federal voting laws and the U.S. Constitution Amendment 14 section (1). The 14th Amendment provides in relevant part that, "No State, shall make or enforce any law which shall abridge the privileges or immunities of citizens of the United States." State laws that allow non-citizens to register to vote and vote in Federal elections abridge the privileges or immunities of citizens of the United States by canceling the effect and sovereign authority of citizens and the citizen electorate.

It is estimated that Biden and Harris have allowed at least 15 million or more illegal immigrants to enter the country since Biden took office. That is in addition to the 30 to 40 million or more that were already here in January of 2021. Many of these people have to pay money to drug and human trafficking cartels to be allowed to cross the Mexican border on the Mexican side. This situation exists because the cartels have significant control of the Mexican border. The cost of crossing the border is estimated to be about $2000 to $3000 per person. This amount is of course a huge sum of money for poor people coming from impoverished areas of the world. Some have money, and some don't. Many of the ones who don't are indebted to the drug cartels or forced to earn money by participating in drug or sex trafficking or by paying back their debt to the drug cartels in remittances of U.S. dollars.

The income of criminal drug and human trafficking organizations has exploded under Joe Biden's administration. Their income went from an estimated $1 billion per year under Trump from 2016 to 2020 to currently about $5 billion or more per year. A House committee report in 2023 concluded that these groups are, "some of the most violent and dangerous groups in the world. These groups spread violence, corruption, and drug-related homicides throughout both Mexico and the United States."

Our neighbor Mexico is on the verge of becoming or already is, a full-fledged narco-state essentially controlled by drug cartels. Biden's policies have facilitated a massive supply of American cash to very dangerous and ruthless criminal organizations. These organizations are obviously aware that billions of dollars in income for them depend on an open, unregulated border and also on Democrats remaining in power. Some of this money is being recycled and used to fund political campaigns of Democrat and globalist politicians. Through the action of some fundraising and political action groups, thousands of contributions amounting to millions of dollars are passed to political candidates using the names and identities of small donors.

In January of 2023, Governor Gregg Abbott of Texas wrote to Biden that the refusal of the Biden/Harris regime to enforce existing Federal law and court orders and the refusal to protect the United States and the several States from invasion constituted a breach of the President's Constitutional duty to execute the law faithfully. The implication of the letter was that Biden's willful refusal to execute his duty to preserve, protect, and defend the sovereignty and the Constitution of the United States did not mean that Texas was prohibited from defending its borders from invasion. Governor Abbott is obligated by the Constitution of Texas to defend Texas' border. Later, there were standoffs between Texas and Federal officials over Texas constructing its own border defenses, which Biden ordered torn down.

Biden and Harris's actions in opening the borders and, in effect, turning control of the borders over to criminal drug and human trafficking cartels are perhaps the most obvious examples of Biden acting in bad faith. Biden's abuse of Presidential authority to aid, abet, and facilitate a literal invasion of the United States is absolutely a breach of the social compact. We are already on the fringes of a post-constitutional era when the executive branch willfully abdicates its responsibilities to defend the national territory.

There is no legitimate reason for open borders and no innocent, good-faith explanation as to why the drug cartels are in substantial control of our border. The costs and risks for the American people and the American Republic are massive. The benefits are non-existent. The beneficiaries of this policy are the Democrat Party and the long list of America's enemies: the drug cartels, Hezbollah, Hamas, other terrorist groups, and the Chinese communist party. Biden and Harris have allowed and facilitated tens of thousands of hostile actors to be present within the United States and be in a position to cause immense damage.

Biden and Harris's refusal to protect the American people from an invasion of military men, terrorists, and criminals and in fact, facilitating the invasion by providing mass transport, flights, and financial support is an act of violence against the American people and the American Republic. Thousands of Americans have already been victimized or died from homicides and drug-related deaths that would not have occurred if Joe Biden had left the border intact and faithfully executed his duty. There is currently a Chinese military and police presence on American soil.

There are other groups as well such as Hezbollah, Hamas, and Venezuelan and Syrian military operatives. It is probably just a matter of time until these forces are called into action by their superiors in Beijing, Tehran, Pyongyang, Langley, or anywhere else.

What will happen if they succeed in transforming the American Republic? If you like the current regime in Venezuela, you will love post-constitutional America. The one-party dictatorship that the Democrats seek to establish if they succeed, will be a very brief reign before America unambiguously enters a post-constitutional era. Once we abandon the very concept that there should be such a thing as national sovereignty or citizenship or free, fair, and auditable elections by a citizen electorate, we will never get any of those things back. Once we destroy the Supreme Court of the United States, the rule of law in the United States will effectively be eliminated. The surviving strains of Constitutional authority will be politicized and subject to mob rule.

What will happen if U.N. bureaucrats overrule Joe Biden or his successor puppets and say we aren't taking in enough immigrants and there are tens of millions more that need homes and care in the area that used to be the United States of America? Joe Biden is already trying to relocate millions of Palestinians to the United States. Who are we, the American citizens, to say no to that?

How will it be possible to defend the national territory militarily after political sovereignty has been surrendered? Why would any American fight the blue helmets or a hostile military or terrorist force on American soil 4 or 6 or 10 years from now (after it is probably too late to do anything about it) if they will not peacefully fight the drug cartels and the Democrats now? The same people, such as Kamala Harris, for instance, who are telling us today that we don't need to be concerned about a massive invasion of illegal immigrants, terrorists, military elements, and criminals will tell us tomorrow that the collapse of the American Republic is actually a good thing. We have finished the job! We have turned the page! They will tell us it is time to embrace the new course—a post-constitutional era. The United States dollar was a wonderful thing while it lasted, but its day is over. It is time for us to let the government control all of our assets, all of the assets of the collective so that every good global citizen can continue to eat and have a place to stay. They will tell us that the U.N. or other military forces on American soil are not hostile forces. They are here to liberate us; to enforce lawful international health and U.N. mandates; to help defeat white supremacy and subversive nativism and to integrate America into the new world order.

In 2019, the Chinese communist party declared a people's war against the United States. The CCP leadership has urged its people to prepare for war. A Congressional committee found that the CCP has infiltrated the United States by coopting elite groups and educational institutions and has been engaged in acts of War against the United States. At least 1 CCP bioweapons lab was discovered in California. The Biden and Harris response has been to surrender billions in weapons and military equipment and an air base that cost hundreds of billions of dollars and allow Chinese military elements to further infiltrate the United States. Not to single China out, there are other groups who are taking advantage of the chaos that has ensued due, in part, to America's long-term path of progressive social reconstruction.

An example of how successful the media has been in normalizing violence against our county and bad faith with respect to immigration was obvious in the border bill debate of February 2024. Joe Biden and Kamala Harris, by their actions, said, "We will secure the border, something we have both the duty and the authority to do right now if Congress gives another $89 billion to the Ukrainian regime and agrees to eliminate the concept of illegal immigration by declaring 1.8 million illegal immigrants per year to be legal." This blatant and obvious attempt to extort the American people to surrender their sovereignty regarding immigration was regarded by the mainstream media as being somehow normal and a good-faith proposal worthy of discussion and compromise. The media has so successfully conditioned the people to accept self-inflicted violence against our county that Kamala Harris was actually taken seriously in the debate when she suggested that the violence and dereliction of duty that she and Joe Biden are responsible for is somehow Donald Trump's fault because he won't surrender our national sovereignty.

We cannot compromise with a party that is acting in bad faith and is actively and willfully seeking the transformation of the American Republic. Joe Biden's and Kamala Harris's actions of turning the U.S. southern border over to the control of criminal drug and human trafficking cartels must never be endorsed by the elements within Congress that are still acting in good faith and upholding their oaths of office.

The important thing at this point is for the American people to recognize what has happened and to refuse to accept this act of physical and political violence against the American Republic. The American people must insist on border security and refuse to be dissuaded by the lies and extortion they are faced with.

CHAPTER THREE
GOOD FAITH

It is critically important to self-government and ordered liberty that those entrusted with political, administrative, or judicial authority act in good faith. If public officials do not act in good faith, then the efficacy of our Constitutional Republic is strained, undermined, and possibly even brought to the point of collapse if the political process is not allowed to function properly to restore leadership that will act in good faith.

Good faith has been defined as:

Good faith is a broad term that's used to encompass honest dealing. Depending on the exact setting, good faith may require an honest belief or purpose, faithful performance of duties, observance of fair dealing standards, or an absence of fraudulent intent.

A fiduciary relationship creates a duty of good faith between the agent and the principal. The breach of this duty of good faith *can lead to <u>liability</u>. Failure to act in good faith is known as <u>bad</u> <u>faith</u> and is generally considered to be a level of <u>culpability</u> greater than <u>negligence</u>.* [1]

A President of the United States certainly has both a fiduciary duty and an actual affirmative duty to act in good faith pursuant to the President's Oath of Office. This duty requires the President to exercise his authority without any fraudulent intent. Good faith is acting honestly, consistent with one's oath to uphold and defend the Constitution of the United States. Where the good faith may end and become bad faith or, in some cases, may metamorphose into crime is difficult to say. We need not define the point precisely to determine that there are many examples of Biden and the deep state regime acting in bad faith. Aiding and abetting the criminal actions of others by a public official is not only bad faith, it is potentially a criminal act in and of itself. Both criminal and bad faith means are being employed to attack and defeat the American Republic.

Much of our contemporary political discourse is based on the premise and assumption, which is uncritically accepted, that Biden and his enablers, including those in the judiciary, are acting

---

[1] Wex Definitions Cornell Legal Institute (Jan. 2023)

in good faith. This is a completely false premise. Biden, Harris, and many others are not acting in good faith.

Many of Bidens enablers are acting with a fraudulent intent that is purposefully hidden from the American people. Part of their intent is to "transform" America, further an agenda of progressive social reconstruction, and ultimately end America's existence as a free, sovereign Constitutional Republic.

Good faith is an important characteristic of a person's actions. By examining critically whether or not someone is acting in good faith we can discern what a person's true motives are. If Biden and his deep state enablers are acting in bad faith or with fraudulent intent, it is reasonable for people to consider their true intent. Their true intent is that free and fair elections by a citizen electorate should eventually be eliminated from the American political process.

The refusal of elected or appointed public officials to act in good faith is itself a breach of the social compact. The refusal to act in good faith is a violation of both the express and implied obligation in law that the President or any official or judge has to act in good faith. A political leader or public official acting pursuant to authority derived from our Constitution and social compact cannot act to undermine and breach our social compact while at the same time seeking its authority and protection. This fraudulent intent is the very essence of bad faith and unclean hands. An example of this form of bad faith is the loud, aggressive citation of the Constitution by hypocrites who want to deny the American people their right to choose the President.

The lack of good faith on the part of Biden and the Democrat Party as a whole is obvious in a wide range of circumstances. Bad faith is evident in the massive lawfare initiated and coordinated by Biden and his enablers against President Trump. Initiating criminal prosecutions for an improper purpose of election interference is not only bad faith it is also a form of prosecutorial misconduct because it is abusing a public trust by initiating a criminal prosecution for an improper purpose namely, election interference.

Once the false premise that the leadership of the Democrat Party is acting in good faith is confronted and recognized, their actions and our appropriate response as American citizens to their actions become much more obvious.

During the debate in June 2024, Biden proudly announced that many of Trump's former associates and cabinet members refused to endorse Trump's re-election. Biden asked why. The reason why is the effectiveness of the Democrats' campaign to intimidate anyone who supports Trump. They are rightly concerned about being charged with crimes, investigated, disbarred, threatened, imprisoned, or having their children and families threatened, as has happened before in many other cases. For the first time in American history, members of Congress who oppose this regime have a well-founded fear of retaliation.

The events of June and July of 2024 and the removal of Biden have made it obvious that there are powerful forces behind the scenes who are working to shape events and control the Presidency. After the debate in June of 2024, it was obvious that Biden was medically unfit to continue the campaign and had no chance to be elected in a legitimate election, a kind of power vacuum emerged within the deep state and the Democrat Party. The power brokers had to be able to select a new President without the risk that someone who could not be controlled by the deep state, like Robert Kennedy Jr, might emerge in a brokered convention.

The simple, undeniable fact that we have a cognitively impaired person sitting in the Oval Office raises a number of disturbing questions. How long had this been going on? Why weren't the American people told of this? If they were lying to us about Biden, what else would they be lying about? Maybe they did lie about the elections of 2020 and 2022. Biden became incapacitated during a trip to Las Vegas or at some point in the summer of 2024. While Biden was still incapacitated, a letter was released purporting to be signed by Biden, saying he would not seek re-election. The deep state was faced with a quandary. If Biden is unfit to run, why is he fit to serve until January 2025? They basically bypassed the 25th Amendment process in which an incapacitated President may be removed without impeachment. Kamala Harris, in effect, replaced Biden as the nominee without any formal process or even getting any votes.

On July 13th, 2024, there was an attempt to kill President Trump in Butler, Pennsylvania. The investigations are just starting but one thing is clear: Either one 20-year-old acting alone nearly killed President Trump from a position 150 yards away, or the deep state is doubling down on its unrestrained lawfare and is now embracing violence to prevent the American people from re-electing Donald Trump. Some of the issues that need to be addressed in any legitimate investigation include: Who ordered the attempt on Trump's life? Why was the attempt made? When was the decision made? Are there elements within the deep state that participated in the planning, execution, or cover-up of the attempt? Would Joe Biden have dropped out if the attempt had succeeded?

It was obvious that Biden was not in control, but who was? The elements and groups behind the scenes that seem to be playing a major role in manipulating events may include Barack Obama and his entourage and allies. These groups could also include current or former Chiefs of Staff and some of the liaisons between the international globalist movement, such as the WEF, various NGOs, U.N. operatives, and the American bureaucratic state.

There is also an important deep-state apparatus called the CIGIE (The Council of Inspectors General for Integrity and Efficiency). The CIGIE has a statutory basis for its existence, however, its role appears to have evolved into that of a deep-state enforcer. Whistleblower reports of illegal activity within the bureaucracy can ultimately be stifled or killed by the CIGIE. Through the process of exercising discretion in investigations and enforcement, favoring some investigations and disfavoring others, the CIGIE and other elements within the deep state can play a significant role in shaping the decisions that ultimately become the policy of the executive branch of government.

The deep state and its media allies are trying to convince the American people that we don't even need a true president anymore—a corrupt dementia patient will do. The deep state can decide how and where the military should be deployed. The CIGIE and top bureaucrats can decide what our national policy should be. The CIGIE and the deep state can determine who will play the role of President. With efficiency and integrity, they can decide who should be investigated, censored, arrested, prosecuted, sued, sent to prison, or whose property should be seized. The CIGIE and top bureaucrats can also decide which Article III judges should be investigated and, if necessary, impeached.

Kamala Harris is the most radical left-wing candidate to be nominated by a major party. As the open borders czar, she literally helped to facilitate the invasion of the United States. Through dereliction of duty, she has played a critical role in the deployment of hostile military elements, terrorists, and criminal gangs within the U.S., not to mention millions of illegal immigrants. She has also been an indispensable part of the 3.5-year-long manipulation of Biden by the Democratic Party and the deep state, hiding his disability from the American people. This fraud on the American people has already been extremely costly and dangerous for the U.S.

The Democrats have already announced their plans if the ballots that are mailed or dropped in drop boxes in America's national election were to grant them political authority in the House of Representatives. They intend to vote to not certify President Trump's election, regardless of the vote or the mandate delivered by the citizen electorate. Could there be a clearer or more vivid statement of bad faith and fraudulent intent than that? The proposition that President Trump be disqualified from serving as President, despite a mandate from the people, is a totally unacceptable usurpation of popular sovereignty. President Trump and all of us were victims of both an uncertifiable and fraudulent election and the Fedsurrection. The President was also acquitted in a trial before the Senate in the second impeachment attempt.

It is critically important that the Democrat/deep state party not control the House of Representatives or the Senate. Republic-loving American citizens should vote for the Republican candidate for House and Senate in their district, even if they don't agree with them on many issues or even know anything about them.

Under Joe Biden's watch as President, we surrendered over $85 billion in arms and military equipment to the Taliban in

Afghanistan. While Biden was vice president under Barack Obama we surrendered about $100 billion in military equipment in Iraq. That withdrawal was at least a bit more orderly. Some of that equipment has already been used against us or our friends and allies. The October 2023 atrocities perpetrated by Hamas and Iran against Israel used equipment that was surrendered in Afghanistan. Is it possible that we could have surrendered $185 billion dollars in military equipment due only to good faith incompetence?

The actions of Biden and his enablers speak much louder than their words. In reference to false prophets, Matthew 7:17 teaches us, "You will recognize them by their fruit. Are grapes gathered from thornbushes or figs from thistles?" Their actions and the results of their actions reveal their true intentions. Their intentions are false. The fruits of their actions have already been nearly catastrophic for the American people.

The actions of the Democrat Party and its allies and enablers including some Republican globalists in financing, facilitating, and advocating the most extreme forms of lawfare and the weaponization of our legal and criminal justice system do not make sense if interpreted as being motivated by a sincere, good faith effort to preserve, protect and defend the Constitution of the United States of America. These same actions make perfect sense if interpreted as being motivated by fraudulent intent, bad faith, a lack of respect for the Constitution, and a literal act of violence against the American Republic.

These actions are designed to consolidate the gains the globalists made in 2020 and ultimately to eliminate citizen sovereignty and the consent of the citizen electorate permanently from the American political process. If the globalist elements won't allow free, fair, and auditable elections by the citizen electorate now why would they ever allow them in the future? If they refuse to act in good faith now when they are fighting for control of the American Republic why should we believe that they will act in good faith if only they succeed in eliminating citizen sovereignty? So-called "Democracy" henceforth will be a charade a farce in which real power will be eviscerated and removed from the hands of American citizens. In a post-constitutional America, there will be elections that are ostensibly legitimate, for a while, but eventually, real citizen sovereignty will be effectively eliminated from our political process.

When the government itself refuses to act in good faith it will not be long until many citizens, non-citizens and illegal immigrants abandon good faith and honest dealing. Top leadership of the Democrat Party as well as media outlets have announced that if Donald Trump is elected there will be a civil war. Really? Why? There is absolutely no reason why there should be a civil war. How about if, instead of a civil war, the leadership of the Democrat Party accepts the will of the people and participates in our social compact peacefully and in good faith for the good of the nation?

Many of the actions by the Democrat Party leadership and its enablers are not taken in good faith. They are not taken to preserve, protect, and defend the constitution of the United States as is required by the President's oath of office. The party's actions as a whole are motivated by a fraudulent purpose—to transform the United States and to further the cause of the color revolution that succeeded in preventing the inauguration of President Trump in 2020. It is almost inconceivable that the end goal of the current Democrat Party is somehow consistent with the responsible, lawful, good-faith maintenance of our Constitutional Republic.

The Democrat Party's actions and plans are inconsistent with both citizen and American sovereignty. There is abundant evidence of their bad faith that should convince the American people that this Democratic party, at least at the national level, is unworthy of the trust of the American people. They have broken the social compact with the American people by among other things facilitating an invasion of our country. They have breached their fiduciary duty to the American people. They have even threatened civil war against the American people if they don't get their way. These are not the actions of unbiased public officials acting in good faith. These are the actions of people who are acting with a fraudulent purpose and are totally undeserving of public trust. If we are foolish enough to allow these obviously malign people to have political authority over sovereign American citizens who will save us from self-destruction?

The extreme duress that the Democrat Party and deep state have subjected President Trump and his supporters to is a direct attack both on our Republic and on all of us. It is a very clear message to the American people. The message is: Don't you dare speak out or associate yourself with Donald Trump or object to the destruction of the American Republic or we will destroy you. We will use lawfare, mobs, the media, criminal prosecutions, and anything we can to destroy you. Those actions and that message are totally inconsistent with good faith participation in our social compact.

The Democrat Party as a whole is not now acting in good faith and has no intention of acting in good faith in the foreseeable future despite the good intentions of many, or even most, of its members. There is no innocent, good-faith explanation for the aggressive actions to subvert the American Republic on so many fronts: The lawfare, the open borders, the willful destruction of our currency, energy supply, economy, and the attacks on free speech.

The lack of good faith by itself should be enough for the American people to conclude that this Democrat Party must be removed from political power at the national level before it causes more damage. There is precious little distance between the current Democrat Party leadership and the Antifa Fascists who make little secret of their desire to burn the Constitution right alongside the American flag. The American people absolutely must respond accordingly before it is too late.

CHAPTER FOUR

ELECTION OF 2020

Our Constitution is our ultimate source of law. The United States Constitution is what separates us from civil chaos and violence. In our Constitutional Republic, the citizen electorate has the ultimate responsibility to vet politicians and determine which ones should be empowered in federal elections and which policies should be pursued. We should not forget the truism that, "The government derives its just powers from the consent of the governed." If elections are fraudulently manipulated, the government does not have the legitimate consent of the governed. Without the consent of the governed, the government cannot exercise its authority in a just manner. It is like the fruit of the poisoned tree. Once the legitimate consent of the governed is terminated, the tree will never bear wholesome fruit.

The years 2016 to 2020 of President Trump's first term had been very good years in many respects. They were years of relative peace and prosperity. As the 2020 election approached, it was clear that the stakes were enormously high. The future of the American Republic was literally at stake. By the count of lawful, legitimate votes, the American people probably chose to reelect Donald Trump by a significant margin.

On the night of Nov 3, 2020, as Joe Biden made up a 659000 vote deficit in Pennsylvania in the middle of the night and massive deficits in the other key battleground states of Georgia, Wisconsin, Michigan, Nevada, and Arizona, it was clear to reasonable, sober and fair-minded people that something was very wrong. These results were not merely statistically highly improbable. The results were statistically impossible.[2]

The basic argument that is made by those who believe the 2020 election was uncertifiable and even fraudulent is that the election had massive irregularities in every critical battleground state. In hindsight, we can see that the result of the political struggle, the "election" of Joe Biden, is not supported by verifiable, auditable votes. The combined effect of these irregularities rendered the reported results unreliable. The reported results

---

[2] See, Pennsylvania Election Analysis by Seth Keshel – US 2020 Election Fraud at a Glance (electionfraud20.org) and other works by Seth Keshel.

were unreliable because they could not be audited and verified. The results, especially in Pennsylvania, Wisconsin, Michigan, Georgia, and Arizona are therefore uncertifiable. The decisions to certify those results in most of the 5 States except Georgia were mainly made by Democrat Secretaries of State, in some cases certifying results that were clearly uncertifiable or not certifiable according to their own State law as happened in Pennsylvania.

A ballot and a vote are not the same thing. When a ballot is counted that does not have a corresponding voter there is a voter deficit meaning the number of ballots or votes exceeds the number of voters credited with participating in the election. These numbers have to be reconciled before the election can be certified. In some states, like Pennsylvania, for example, the results cannot be lawfully certified if these numbers cannot be reconciled.

In order to create fraudulent "votes" vote traffickers need several things. They need blank ballots. They need names from the voting rolls that can be applied to mail-in or drop-box absentee ballots such as the names of dead people, those unlikely to vote, or those who have moved out of State but are still on voter rolls. Or, they need the name of a ghost voter i.e., a fictitious person who has been placed onto the voter rolls in order to obtain a ballot, which can then be submitted as a "vote". The work of many authors and organizations, such as True the Vote, Verity Vote, Dinesh D'Souza, Patrick Colbeck, and others, have documented the operations of vote traffickers.

Election law is a mixture of both State and Federal law. Each State has its own election laws but there are also Federal laws applicable to elections for Federal office holders such as the President, House, and Senate. Federal election law requires certain standards in terms of maintaining voter rolls that are consistent with that state's living and qualified voters. These qualifications are minimal, that the person be alive, that they be a citizen, and that they are a resident of that State. States are required, pursuant to Federal law to cull the voter rolls to eliminate ineligible people such as dead people, incompetent people, and those that no longer live in the State. One of the purposes of Federal voting laws is that the voting rolls be accurate and there be a precise number of eligible voters. If the number of "votes" exceeds the number of voters that is an indication that something may be wrong. Another aspect of Federal election law is that the acceptable error rate is within .08%.

Common problems in most or all the results in the battleground States included: large numbers of absentee ballots with unsigned or unverifiable signatures on the ballot exterior envelope; the inability in some States to reconcile the number of ballots counted as votes with the number of registered voters who cast ballots; ballot drop boxes being deployed without sufficient, or any, lawful authorization; ballot harvesting in some cases contrary to State law; unprecedented large volumes of absentee ballots; lost or non-existent records documenting the collection and transfer of ballots from drop boxes; large numbers of ghost voters; at least some numbers of non-citizens registered to vote and some of them casting ballots; evidence of Republican votes being removed and absurdly high percentages of votes for Biden in some areas.

There are numerous statistical anomalies as well. In 2020, Republicans gained 14 seats in the House. The gain was the largest gain by Republicans in an election cycle since 1942. To believe that Biden got 81 million votes, the most of any Presidential candidate ever, and the Republican party still gained 14 seats in the 2020 Congressional House election, by implication it has to be concluded that tens of thousands of Republican voters in key swing states voted for the down-ballot Republican candidates but refused to vote for President Trump.

In hindsight we can see that the end result of the political struggle, the "election" of Joe Biden is not supported by verifiable, auditable votes. The combined effect of these irregularities rendered the reported results unreliable. The decisions to certify those results in many of those States, with some exceptions, were made by Democrat Secretaries of State, not the State legislatures as provided in Article 2 sec. 2 of the U.S. Constitution. In some cases, results were certified that are not certifiable according to Federal Law. In some cases, results were certified that were not certifiable according to both State and Federal law as happened in Pennsylvania for example.

President Trump published a report on one of his websites entitled, "Summary of Election Fraud in the 2020 Presidential Election in the Swing States". The report is readily available online and is extensively documented.[3] There are numerous other sites that have evidence about the 2020 election. The report has been criticized for various reasons including that it is unsigned. The report cites numerous examples in each swing State of irregularities and problems in that State that were enough to change the outcome of the election.

---

[3] Battleground states 2020 election fraud claims.pdf - Google Drive

Pennsylvania

According to the report, in Pennsylvania, there were 7,035,746 ballots counted as votes. There were 6,914,556 voters credited with participating in the 2020 election. This means there were at least 121,190 votes counted that did not have an identifiable, qualified voter.[4] The number of voters who participated in the election should be equal to or very nearly equal to the number of ballots received and counted. If a voter goes to their local precinct and shows their ID (if required in that State), election officials check the register, and if they are registered, issue the voter a ballot and the voter is credited with having participated in the election. Or if they mail in a ballot, their name is checked off the list of those registered to vote who have received a mail-in ballot and if the signature matches the control signature on file, their ballot is counted and they are credited with having participated in the election.

Another serious problem was the inclusion of ballots in Pennsylvania that were received after November 3rd, 2020. Before the election, there was already litigation regarding supposed changes to Pennsylvania's election laws that would,

---

[4] Verity Vote, 'Pennsylvania Voter Deficit', Feb. 10, 2021, https://verityvote.us/pennsylvania-voter-deficit

among other things, allow postmarked ballots to be received after November 3rd. Eventually, it appears even non-post marked ballots were received. Pursuant to some of this litigation, ballots received after 3rd November were supposed to be sequestered. According to the president's report, there were 71,893 ballots received after November 3rd with some ballots received as late as November 12th. These numbers were apparently included in the totals of those credited with participating by mail. According to the Secretary of State, there were only 10,000 ballots received between November 3rd and November 6th. How these ballots were accounted for and if they were even lawful under Pennsylvania law appears to be a matter of some legitimate dispute.

Another serious problem was the unlawful counting of thousands of mail-in ballots without Republican poll watchers allowed to attend and participate. This was done in Philadelphia despite a court order.

In Pennsylvania, there were numerous cases of people going to the polls to vote being told they had already voted. Apparently, their name was used and put on a mail-in ballot because they were on a database of voters unlikely to vote.

Pennsylvania had some of the most prolific activity of vote traffickers in the United States in 2020. True the Vote, a voter rights advocacy group, used data analysis of cell phone data to identify individual cell phones that had made at least 10 trips to ballot drop boxes and 5 trips to non-governmental organizations in the time frame around November 3rd 2020. They found 1,155 cell phones fit this criterion. This information is additional circumstantial evidence that any voter deficit or the massive surge in ballots for Biden in Pennsylvania may have been caused in part by the operations of vote traffickers.

These were some of the issues that led to serious concern by the Pennsylvania legislature that the certification by the Secretary of States Office may not have been valid. The voter deficit and other identified problems exceeded the margin of victory for Joe Biden. The margin of victory for Biden was 80,555 votes. Without Pennsylvania, Joe Biden loses the election. There were many other irregularities as well in Pennsylvania that rendered the results uncertifiable.

One of the U.S. Attorneys in Pennsylvania was instructed by the U.S. Attorney General not to investigate allegations of election irregularities and to refer any "serious" matters to the partisan

Pennsylvania State Attorney General who would of course not do anything about them.

Arizona

The margin of Biden's victory in Arizona was 10,457 votes. The Summary that was released claims that there were over 20,500 ballots that were received and counted in Maricopa County, AZ after the deadline of November 3rd at 7:00 PM. Arizona law requires ballots to be actually received or deposited at a polling place by Nov. 3rd. Of the 20,500 ballots, there were 18,000 that have been the subject of some controversy. One of the sources cited in the Summary is a report by Verity Vote. The Verity Vote report outlines several interesting points. First, the documents authenticating the chain of custody of these 18000 ballots were not released by Maricopa County for over 7 months, long after January of 2021. [5] The report contains information that there were 2 entries on the report. One entry indicated there were 17,150 ballots transferred from MC (central computing location) and delivered to the 3rd party contractor who scanned the ballots. These 17,150 ballots are not in question. There is also

---

[5] 'Long Withheld Records Reveal More than 20,000 Mail Ballots Received After the Legal Deadline', Verity Vote, 2022, https://verityvote.us/long-withheld-records-reveal-more-than-20000-mail-ballots received-after-the-legal-deadline/

a second entry on the ballot custody receipt signed by 2 election officials dated November 4th, 2020 that shows 18,000 ballots listed as USPS inbound. Meaning, that these 18,000 ballots were not in Maricopa County custody as of 7:00 PM November 3, 2020, according to the witnesses that signed the ballot custody form.

There were many other problems in Arizona as well. Arizona had a high number of ghost voters. The existence of these ghost voters has been documented by independent surveys using canvassing data.

There were also significant issues with the signature verification of mail-in ballots. There are standards about what constitutes a "signature" according to the Arizona Secretary of State. Unsigned envelopes are not signatures. A signature of a different person other than the voter is not a signature. A signature unreasonably different than the control signature on file is not a signature. In Arizona, there were 1.9 million mail-in ballots. Evidence was presented to the Arizona Legislature that there were at least 76,364 mail-in ballots in which there were egregious signature mismatches. The true number is probably much higher than that.

Georgia

In Georgia, Biden's victory was by a margin of 11,779 votes. There were numerous problems with the results in Georgia. Georgia was a particularly important battleground state because of the unusual circumstance of 2 U.S. Senate seats, both seats were held by Republicans, being up for reelection.

For starters, there were 364,000 ineligible voter registrations on the rolls in 2020. It appears there were about 67,284 ineligible voters who cast votes. These people could have been illegal immigrants, non-citizens, dead people, people registered in other States, or otherwise ineligible persons.

In Fulton County GA, most of Biden's votes were absentee or mail-in ballots. According to the report, citing Voter GA materials, approximately 90% of 148,000 absentee ballots cast in Fulton County cannot be authenticated. Moreover, 132, 284 mail-in ballots have no electronic authentication file. [6] When mail-in ballots are opened at a polling place or collection center, ideally there are poll workers from each party there. The name of the

[6] 'VoterGA Press Conference March 7th', Rumble, 2022, https://rumble.com/vwmwupvoterga-press-conference-march-7th.html (30:38)

sender is checked against the voter rolls. In theory, the signature is also checked. If the ballot is determined to be authentic, the sender's name is checked off and that person is credited with voting and a cast vote record is created. The outer envelope is separated from the ballot inside. Depending on local rules, the outer envelope may also be scanned. The ballot is scanned and a digital image of the ballot is created. When the ballot is scanned, the tabulator produces both a digital image of the ballot and an electronic authentication file. Each tabulator has a log of how many ballots were scanned on that tabulator. Each scanned ballot should have an authentication file. The tabulator can then print a tape that documents the number of ballots scanned on that tabulator. According to Georgia law, the poll manager and 2 poll workers have to sign off on the tape to authenticate that they witnessed the scanning of the ballots. In theory, the number of actual ballots should equal the number of authenticated scanned ballots. Moreover, there should be a cast vote record that identifies the registered voter who participated in the election as opposed to a registered voter who may not have participated. The number of voters credited with participating in the election should equal the number of votes within .08% and there should be an authentication file for each scanned vote. If there are large numbers of ballots that cannot be authenticated, it is an indication of a serious problem at the very least.

Statewide, there were 235,000 absentee ballots requested and accepted before the earliest date for the 2020 election which was May 6th, 2020. In Fulton County, there were 315,000 early votes. According to the report, none of these votes were properly witnessed and attested to by poll workers. [7] "The closing tapes for these votes are all unsigned, showed more tabulated votes than the tabulators had recorded as scanning in their protective counters, and recorded improbably low percentages for President Trump. . . . The anomalies indicate ballots were not scanned on the tabulators that printed the closing tapes, making the closing tapes fraudulent." (Summary of Election Fraud in the 2020 Presidential Election in the Swing States, p. 7).

Wisconsin

In Wisconsin, the margin of victory for Biden was 20, 682 votes. In Wisconsin, there were at least 3 major problems with the 2020 elections. The first problem was the illegality of the drop boxes that were used all over the State and the lack of any video surveillance of the drop boxes. The second problem was the use of the circumstances of COVID to relax voter ID requirements and the third was an unprecedented surge in votes from "indefinitely

---

[7] 'Unsigned Tabulator Tapes in Fulton County - Nov 2020', Rumble, 2022, https://rumble.com/vz5keh- 15 unsigned-tabulator-tapes-in-fulton-county-nov-2020.html

confined persons" who normally might need the assistance of special voting deputies (SVD) to cast a ballot because of the very high incidence of incompetence among the population of indefinitely confined persons. The increase in participation of indefinitely confined persons averaged almost 300% statewide and almost 500% in Racine County.

In Wisconsin, neither the Wisconsin Election Commission nor the Legislature had ever voted to allow drop boxes for use in the 2020 election. One single commissioner of the WEC decided drop boxes could be used. Eventually, the Wisconsin Supreme Court in July of 2022 decided that drop boxes were not legal under Wisconsin law. In Wisconsin, there were 528 drop boxes deployed Statewide. In Wisconsin, there were only 653,236 votes cast in person. There were 1,969,274 absentee and mail-in ballots cast. A large percentage of the ballots were from drop boxes including 217, 424 in Milwaukee alone.

Voter ID requirements were relaxed by local election clerks and officials "ruling" that Wisconsin voters were "indefinitely confined" due to COVID-19 and therefore were not subject to photo identification requirements to obtain an absentee ballot. This means that people could obtain an absentee ballot without having to upload or present a photo ID. By December of 2020, the

Supreme Court of Wisconsin had ruled that COVID did not render Wisconsin voters indefinitely confined and there was no lawful basis to issue absentee ballots without photo ID.

Michigan

In Michigan, 5,579,317 ballots were cast and certified. Michigan's qualified voter file for Nov. 3, 2020, has never shown there were 5,579,317 people registered to vote in Michigan as of November 3, 2020. [8] The highest number of qualified voters ever reported by the Michigan Secretary of State for November 3, 2020 was 5,511,303. This means there is a voter deficit on paper of at least 68,014.

Even if 100% of the registered voters actually voted, there could not have been more than 5,511,303 lawful votes cast. According to the Summary Report, the list or number of voters credited with voting in Michigan has been in a state of flux. In December of 2023, there was a total of 5,307,751 listed as having voted in the 2020 election. This number would imply a voter deficit of approximately 271,566 which was 1.5 times the margin of Joe Biden's victory. A list of the number of voters credited with

---

[8] Tim Vetter, 'Michigan's Voter Roll History Data Manipulation', CheckMyVote.org, December 2023

voting in Michigan has never been released. The actual number of the voter deficit could be more than 271, 566. For example, if 95% of qualified voters actually voted instead of 100% the maximum number of lawful votes would be 5,511,308 x .95=5,235,742. If 5,579,317 ballots were certified and the maximum number of lawful votes is 5,235,742 the voter deficit would be 5,579,317-5,235,317= 343,575.

In Michigan, thousands of ballots were brought to the central counting facility in Detroit at 3:30 AM in the middle of the night 8 hours after the deadline. Moreover, there was no legitimate chain of custody documentation for these ballots. This event is documented by multiple affidavits and video evidence.[9] Critics of this evidence claim that the ballots were deposited in drop boxes by the 8:00 PM deadline and had simply not arrived at the collection center yet. This controversy illustrates some of the problems both with drop boxes and with counting ballots that do not have a chain of custody documentation. There is no way to verify that the ballots were deposited by 8:00 PM. When ballots are collected from a drop box, they are supposed to be collected by election officials, ideally a Republican and a Democrat. The

---

[9] Exclusive: The TCF Center Election Fraud - Newly Discovered Video Shows Late Night Deliveries of Tens of Thousands of Illegal Ballots 8 Hours After Deadline | The Gateway Pundit | by Jim Hoft

chain of custody documentation requires 2 signatures. In other words, 2 known identified election officials have to certify that they witnessed the collection of these ballots before 8:00 PM on November 3, 2020. Apparently, that did not happen in this case. There had already been a ballot drop of 138,000 ballots at 10:30 PM. At 3:30 AM, Biden was behind by well over 100,000 votes. By about 7:00 AM on November 4, 2020, Biden had pulled into the lead. It is a reasonable conclusion from the evidence that the votes from Detroit that arrived at 3:30 AM must have had a very high percentage for Biden and also contributed significantly to Biden going into the lead.

These problems are an example of what happens when ballot security measures are dismantled and ignored. The end result is elections that are unverifiable because they cannot be reproduced or authenticated. Where is the chain of custody documentation and the voters to support these election results? They cannot be identified.

The American people sensed that something was terribly wrong—despite concerted media suppression and a willful, ongoing refusal by both the mainstream and controlled opposition media, such as Fox News, to report any of the evidence of the 2020 election to the American people in a meaningful and coherent way. The American people intuitively understood that a man who couldn't draw a crowd of 250 unpaid people did not get 81 million legitimate votes. Joe Biden did not receive a higher percentage of votes from the predominantly African American communities of Philadelphia than Barack Obama did. We sensed that there had been a *coup d'état* and that we were, in fact, losing our Republic.

The progressive globalist front had made a bold, quasi-legal, and extralegal move. The quasi-legal part involved a massive effort, with unlimited funding from both dark money and legitimate sources, to dismantle ballot security measures, constant lawfare to fight the updating of voter registration rolls and expand mail-in balloting. The illegal part was old-fashioned vote manufacturing and ensuring just the right number of ballots were delivered. If caught and exposed, they risked not only criminal prosecution, potentially for sedition but the progressive globalist vehicle—the Democrat Party itself—might be so tarnished that it could no longer effectively masquerade as an American political party acting in good faith. The party's fraudulent intentions would be so obvious they could no longer be hidden from the people.

In an environment of free and fair elections by an informed citizen electorate, the current Democrat Party could become unable to thwart the will of the American people to protect American sovereignty and the American Republic. But if they succeeded, the pot of gold at the end of the rainbow was right there—the de facto establishment of a one-party globalist/totalitarian state within America. The borders could then be opened, and tens of millions of new Democrats brought into the country, supported at crippling public expense, to assist in preparing America to welcome the new world order. The ties between global woke capital, non-government organizations, the U.N., and a weaponized bureaucracy in the United States could be strengthened. So-called "laws" could be passed, criminalizing free speech, opposition to government corruption, and opposition to election fraud.

Extremists who object to the destruction of the American Republic could be arrested *en masse* and perhaps given an opportunity for re-education, as some leading Democrat politicians have suggested. The progressive globalists gambled, correctly in hindsight, that many powerful Republican officials would be too weak, corrupt, selfish, cowardly, or incompetent to even grasp the seriousness of the crisis the American Republic faces, let alone fight against it. Perhaps some were bribed or blackmailed. Perhaps they were simply not ready to believe that something like this could happen in America. Or, at a minimum, Republican officials would not be willing to take the bold actions necessary to defeat this long-term, willful attack on the American Republic.

The conclusion among informed and fair-minded people was clear—something was terribly wrong with the American political system. The American people in fact had not elected Joe Biden to be President of the United States and yet here he was in the White House. The implications of this simple and obvious conclusion were both disturbing and terrifying—we were losing our Republic and if this corruption wasn't significantly rooted out soon, we would lose our Republic forever.

In the litigation and certification process that followed the 2020 election, the Democrat Party/deep state was extremely successful in transforming what should have been a fact-based issue—whether the ballots counted, especially in critical areas, were legal votes cast by lawful voters—into a political issue. Instead of a fact-based investigation, including evidentiary hearings or trials, the election became a political contest. There were no trials. The focus of post-election evaluation became primarily political, rather than an evidence-based investigation.

In a political contest, the divisions within the Republican Party could be exploited. Yes, there were recounts and audits in some states, but these recounts mostly counted ballots without critically examining whether each ballot was lawfully cast.

When looking at the circumstances of both 2020 itself and the aftermath of the election, as well as subsequent events in 2022, several key facts stand out:

1. The 2020 election was fraught with numerous, easily avoidable difficulties that rendered the results unreliable because the results in critical areas are unauditable and unverifiable.
2. The Democrat Party, with support from certain elements within the Republican Party, did not make a good-faith effort to learn the truth about the 2020 election so that America will never suffer another crisis like that one.
3. These same elements have continually fought election security measures and continue to do so to this day. They are also aggressively seeking to expand voting rights to illegal immigrants and non-citizens.
4. Instead of cooperating to institute reasonable and sound election security measures like those proposed by the Carter Center, the deep state elements, led and orchestrated by the

Biden administration, are using increasingly aggressive lawfare and abuse of the criminal justice system to attack people who acted in good faith to bring these issues to the attention of the American people so they could be effectively addressed.

More evidence about the 2020 election came to light after 2022 and shed additional light on the 2020 election. The allegations that the 2020 election had been manipulated are more credible in hindsight.

The Wisconsin legislature commissioned a study on the 2020 election which was produced by Justice Michael Gableman a former justice of the Wisconsin Supreme Court. In March of 2022, the report concluded that there were numerous irregularities including the illegality of the drop boxes, massive private funding, and involvement of private entities in the election process. Moreover, unlawful conduct by Wisconsin election officials cast grave doubt on the certification of the Presidential vote in Wisconsin.

Virtually every major political candidate, except for Texas Attorney General Ken Paxton, who vociferously advocated for election security measures was vigorously attacked and went down to "defeat". In some cases, their opponents received massive campaign funding. Examples of this include: The attacks in the election against Tina Peters running for the secretary of State of Colorado and subsequent criminal prosecution; the defeat of David Purdue both in his Senate race and subsequent race for Secretary of State of Georgia following his road to Damascus conversion to favor ballot security measures; the defeat of Matt DePerno as Attorney General and Christine Caramo as Secretary of State in Michigan; the defeat of Kari Lake as Governor of Arizona; the defeat of Mark Finchem as Secretary of State of Arizona; the defeat of Abe Hamadeh as Attorney General of Arizona and the installation of a former drug cartel lawyer as Attorney General of Arizona.

The Democrat party knows there is no way they can win elections at the national level long term without the support of massive numbers of non-citizens and illegal immigrants who are not allowed to legally vote in Federal elections. That is the obvious reason why the Democrat party moved heaven and earth to stop the construction of the border wall. They do not want a secure border. Open borders is their policy. There is already reliable evidence that non-citizens voted in the 2020 election.

The rate of infiltration of the border has slowed some because of the upcoming 2024 election but there are millions more illegal immigrants currently being housed in Southern Mexico who will cross the U.S. border after November 5th regardless of who wins the election.

We can already see the plans to subvert the 2024 election are in full swing. Illegal immigrants and non-citizens are being registered to vote in significant numbers. The main source of illegal votes will likely be from the 45 million or more non-citizens some here legally some here illegally who are currently residents in the United States. There have been some 15 million or more illegal immigrants who have entered the United States just under the Biden-Harris regime. The obvious and main purpose these people were allowed into the country was so that they could be given ballots and registered to vote.

In 2024, the Fulton County Georgia Board of Elections enacted a reasonable, common sense measure to require reconciliation of ballot totals and voters at the precinct level before transmission of the data to other entities. The measure did not require any formal recount or vote tabulation. The Democrat party has pulled out all the stops in fighting the measure.

So, the United States will likely again be faced with the unfair situation in which non-citizens and illegal immigrants who cannot legally vote in Federal elections will be registered to vote and handed or mailed Federal ballots in many States. If they cast the ballot in a federal election, it will be both an illegal vote and a crime. This process is not a free and fair election process because it will be virtually impossible to cull out illegal votes prior to them being cast. And, of course, that is the idea.

The solution to this mess could start with the enactment of the SAVE Act which would require States to verify eligibility to vote i.e. American citizenship before actually registering people to vote. Obviously, that can only happen if Democrats are not in control of the legislative and executive branches of government.

CHAPTER FIVE

WEAPONIZATION OF JUSTICE

When Joe Biden said that Donald Trump must not be allowed to become President he meant it. Why? Because Donald Trump will not take orders from the deep state and the military-industrial complex. Worse yet, Donald Trump wants the American people to regain control of their wealth and currency. He wants to take the nation's credit card taken away from the deep state and the math deniers. The election of Donald Trump in 2024 could mean major setbacks for the deep state, serious investigations, revealing secrets, and many of the crimes against the American people that had led to Joe Biden's installation.

A few days after the midterm elections in November of 2022 in which the outcome was that the Republican party emerged with about a 5-seat advantage and therefore would control the House of Representatives in the next Congress, Donald Trump announced he was running for reelection. In

November 2022, President Biden announced he would use "constitutional" means to ensure that Donald Trump would never become President—not that he would defeat Donald Trump in an election. What did that mean exactly? Within a couple of days, the American people and the world found out. There would be a massive lawfare campaign launched on many fronts to destroy Donald Trump and anyone associated with him. Even though it may mean the end of America's existence as a free Constitutional Republic, Trump had to be stopped.

The timing of this lawfare offensive, that it came right after Trump announced he was running for reelection, is a clear indication that the lawfare is motivated by an improper purpose, namely election interference. The weaponization of the Department of Justice, the use of lawfare on a massive scale, and the merciless persecution of President Trump and his supporters put the lie to any claim that the Democrat party or the Justice Department in the current administration is acting in good faith.

America's two-tier system of justice is on full display, out in the open and totally unashamed. On one tier of America's justice system, regime opponents are persecuted and even imprisoned. No target was more important than Donald Trump. On the other hand, regime insiders are coddled and protected by the media, law enforcement, and prosecutors. The obscene corruption of Joe Biden and some of his family taking in millions of dollars from several sources including some Ukrainian oligarchs and entities controlled by the Chinese communist party, was not only ignored by the mainstream media and law enforcement it was actively covered up.

The legal system both civil and criminal is a very important tool that is being used to suppress dissent and to punish and intimidate regime opponents. This abuse of our legal system through lawfare is being deployed both offensively and defensively. The offensive applications are clearly evident in the many attacks on Donald Trump and his key supporters.

The first salvo in a sustained lawfare campaign and partisan abuse of the criminal prosecutorial function started with the persecution of General Michael Flynn right at the beginning of the Trump administration in 2017. General Flynn had just been nominated to be national security advisor to the President. But, there was a big problem. Flynn was already known to be a person who was loyal to the President and who would not cooperate with the deep state. Flynn also was very informed about how the deep state and FBI had illegally manipulated the FISA courts to obtain warrants to tap Donald Trump's telephone. Biden was right there at ground zero, literally in the room, when the deep state initiated a criminal prosecution against Flynn in order to remove him from office.

Starting in late 2022 and into 2023, President Trump was also charged in 2 different Federal criminal proceedings.

The lawfare and weaponization of civil and criminal legal processes continued and intensified throughout 2022, 2023 and 2024. There are several examples of prosecutions that illustrate the extreme bad faith and very dangerous abuse of our judicial system that is being orchestrated and supported by the Democrat party and in some cases supported by their rino allies. Special laws were passed in New York so that President Trump could be civilly sued for alleged offenses that had been time barred for years by the Statute of Limitations. Other attacks against the American people include: The investigation and impeachment of the Attorney General of Texas Ken Paxton; the criminal proceedings against the so called fake electors; The criminal prosecution of Tina Peters the country clerk of Mesa county Colorado motivated by her compliance with federal and state laws regarding the retention of election records; the criminal, civil and bar proceedings against Rudi Guliani; the civil lawsuits against Kari Lake and the widespread attacks on virtually all of President Trumps attorneys using ethical complaints.

The numerous civil and criminal proceedings that have been brought against President Trump are not in any way an indication that he has done anything wrong. The systematic persecutions of President Trump and his supporters are the canary in the mineshaft, a symptom of something far more serious—the weaponization and breakdown of our system of justice. This persecution of the President is not just an attack on one man, Donald Trump. These prosecutions are initiated totally in bad faith and in breach of the public trust. All of these prosecutions are an attack on our system of justice; on our Republic and ultimately on all of us.

A prosecutor under the American Bar Association rules as well as applicable State and Federal rules has numerous ethical obligations with respect to the initiation of a criminal prosecution. Generally, these rules provide that prosecutors are exercising a public trust for the purpose of protecting the public safety interests. Prosecutors should avoid even the appearance of impropriety or that a prosecution has been initiated for an improper purpose. American Bar Association Standard 3-1.2(b) provides;

b) The primary duty of the prosecutor is to seek justice within the bounds of the law, not merely to convict. The prosecutor serves the public interest and should act with integrity and balanced judgment to increase public safety both by pursuing appropriate criminal charges of appropriate severity, and by exercising discretion to not pursue criminal charges in appropriate circumstances. The prosecutor should seek to protect the innocent and convict the guilty, consider the interests of victims and witnesses, and respect the Constitutional and legal rights of all persons, including suspects and defendants.

Criminal prosecutions cannot be initiated in bad faith or for partisan political purposes. To do so is a breach of the public trust and a serious misuse of public authority. Initiating a criminal prosecution for an improper purpose is a violation of ethical rules, which are imposed by State Supreme Courts and by the Federal Rules. Initiating an improper criminal prosecution is also a violation of that prosecutors' oath of office both as a public official and as an officer of the judicial branch of government. In some States, the attorneys Oath contains much stronger prohibitions against unethical conduct than the applicable rules of professional conduct contain.

The prestitute corps and of course big tech dutifully explained to the American people that these prosecutions were just the normal operations of law enforcement. The true purpose of politicizing the prosecutorial function to attack President

Trump and his supporters is to deny all of us the right to choose our President and to attack the American Republic itself. Another purpose is to intimidate citizens involved in reasonable and lawful speech and political activity.

This lawfare is election interference that is being orchestrated by the Biden Justice department and cooperating State level district Attorneys who are willing to compromise their ethical duty to the American people and the people of their State in order to defeat the ability of the American people to choose their President.

The Russian people had experienced a law enforcement regime similar to the Biden regime under Laverentiy Beria head of the Soviet secret police who said, "Show me the man and I will show you the crime." This is literally where the Democrat party and Joe Biden's controllers are taking us.

Vladimir Putin, in an address in September of 2023, said that the ongoing political persecutions in America were a good thing because the persecutions exposed the rotten core of Americas political system. Americas legal and political system was totally corrupt. America could no longer credibly criticize other countries for corruption and failing to uphold the rule of law.

Birth of the Stolen Classified Documents Hoax

A special prosecutor was "appointed" to begin Federal Prosecutions of Donald Trump. The real purpose of the appointment appears to be election interference and to find something, anything they possibly could to convict Donald Trump. A special prosecutor who specialized in politically motivated, unethical prosecutions that could not survive appeals indicted the President in the summer of 2023. The special Federal prosecutor eventually focused on 2 areas of attack: The classified documents hoax case out of Florida and the Fedsurrection case out of DC.

On January 19, 2021, the day before President Trump's term was ended, the President declassified numerous documents. Some of these documents reportedly dealt with the origins of the crossfire hurricane operation. In that operation, deep state elements sought to spy on, subvert and undermine President Trump both as a means to prevent his election and as an "insurance policy", in the event he was to be elected, to enable the deep state to harass and undermine his Presidency. This effort ultimately caused enormous damage and expense to the United States. President Trump did not release these documents even though arguably he could. He possessed them pursuant to the Presidential Records act and began a process to declassify them without necessarily agreeing that the bureaucracy has the ability legally to impair the Presidents authority to declassify documents. Declassification of at least of these documents is ultimately an executive branch responsibility. It is clearly in the public interest that documents related to internal sabotage of the United States and of our political process be publicly revealed.

The President and his team were working with staff at the National Archive to affect the declassification process. The problem for the deep state was that these documents reportedly reveal the involvement of high-level Democrat party officials and their allies in the deep state and intelligence community colluding with each other in the origins of the Russian collusion hoax that began in 2015 and 2016. The basic outline of the hoax is now well known through the efforts of the Devon Nunez committee and others. The Hillary campaign had paid for the fabricated Steele dossier which was then used as "evidence" by cooperating elements within the FBI to obtain warrants from the FISA court to tap into the of then candidate Trumps phone.

There is still much more to be revealed. Particularly important is the process of how known fabricated evidence was presented by FBI and NSA officials to a FISA court as being authentic for the purpose of obtaining warrants to tap Trumps phones—not just once but 3 or 4 times. How much public money was spent on this hoax and on whose authority? This hoax was also a massive in-kind campaign contribution from the Federal treasury to the Hillary Clinton campaign. The hoax eventually metastasized into the Muller commission and impeachment #1 and cost the American people tens of millions of dollars and caused major disruptions in the functioning of the government. One FBI agent so far has been given a slap on the wrist for fabricating evidence in the creation of the hoax but much more work needs to be done to bring some of the perpetrators to justice.

The purpose of the classified documents prosecution had nothing whatever to do with a legitimate, good faith protection of the public safety interest. President Trump and his staff had been cooperating with archivists. The FBI had been to Mara Lago in the spring of 2023 and reviewed the security measures for the documents and had no significant objections other than installing some new locks. This prosecution was also designed to keep these documents under seal because they were evidence in an ongoing "case". It was critically important to the Democrat party and its ability to install another deep state puppet into the White House in 2024 that no more evidence of this massive crime against Donald Trump and the American people be publicly revealed before the election. Another purpose was to smear Donald Trump; undermine his political base; wear Trump down financially and emotionally; impair his ability to campaign and to serve as a warning to any others who would dare to stand up to the deep state as effectively as Donald Trump has.

The issue of whether the Attorney General of the United States even had the statutory authority to appoint a special prosecutor without Senate confirmation was eventually decided by a Federal Court in Florida in the spring of 2024. The answer was no. It is the responsibility of the Attorney General alone or his subordinates to bring Federal indictments. That biding court order however did not stop the special prosecutor from issuing a superseding indictment against Trump after the classified documents hoax case was dismissed.

The so called 'fake' electors were prosecuted criminally. The electors themselves did nothing wrong. The criminal prosecutions of people making a good faith effort to serve their civic duty is a very serious abuse of the prosecutorial process. The obvious goal of this prosecution is to intimidate anyone who actively opposes Americas transformation into a post Constitutional regime. This intimidation is much more disturbing than a mere abuse of prosecutorial discretion.

State legislatures have the ultimate authority pursuant to Article 2 sec. 2 of the Constitution to determine both the certification of elections and the allocation of that States electors. State legislatures are responsible under the Constitution for naming a slate of electors to cast that States electoral votes. The purpose of naming an alternate slate of electors was to prepare for the possibility that the authorization for electors, that had been previously authorized by a Secretary of State, was rescinded by the State Legislature and another slate of electors authorized. Such an outcome was a theoretical possibility. This response of having alternate electors was a perfectly reasonable, good faith, nonviolent and appropriate response to the fraud that was obvious in the 2020 election. After all, it was the Democrats who decided the election should be decided by extreme lawfare rather than a free, fair and verifiable election process. Alternative electors had been named by President Kennedy in 1960.

President Trump was also criminally prosecuted in Georgia for making a phone call to the Georgia Secretary of State.

As President of the United States, Trump has an obligation to do whatever he can to eliminate vote fraud and unreliable election processes whether or not he personally benefits from the elimination of vote fraud. This obligation is both express by his oath of office and implicit by the principles of good faith. By the time that the President's alleged criminal conversations with the Secretary of State of Georgia took place, it was clear that the Attorney General of the United States was not seriously pursuing investigations of vote fraud that may have occurred and in fact was suppressing Federal investigations into vote fraud in Pennsylvania.

The Attorney General of New York had run on a platform to get Trump. This in and of itself is probably unethical. The Biden administration lent its support to these efforts in both the New York and Georgia State cases. A top Department of Justice official left a very powerful position in Washington to be the field general on the New York front in the effort to destroy Trump and put him in prison. The White house counsel's office was also involved in supporting and helping to orchestrate the lawfare in both Georgia and New York.

This lawfare is broad based and goes far beyond just the attacks on President Trump. There is lawfare against the

Presidents allies and attorneys. The top targets are those who are at the forefront. The Democrat Party through the deep state has initiated baseless disbarment proceedings against several of the Presidents attorneys including Jeff Clark and Rudi Guliani. There have also been proceedings against Matt DePerno an election integrity advocate who ran for Attorney General of Michigan. These are serious matters because it means the contention that the Democrats are acting in good faith and acting in the public interest pursuant to legitimate public authority is a lie.

By the spring of 2024 the lawfare had gotten so intense that the Biden regime actually did jail two of the Presidents key advisors, Peter Navarro and Steve Bannon. There have been criminal proceedings against Tina Peters, county clerk in Mesa County Colorado for following federal and State law to preserve election records. No one says any of these people are actually wrong in their allegations that there were significant irregularities in the 2020 and 2022 elections that should be taken seriously and investigated. No, they say they are subversives and purveyors of misinformation who must be punished and silenced.

A second purpose of using criminal prosecutions against not only Trump but those who have stood with the President particularly his attorneys and advisors after 2020, is to terrorize the opposition and to intimidate anyone who dares to demand free, fair and auditable elections or object to the transformation and destruction of our Constitutional Republic. Widespread abuse of the civil and criminal justice system will dissuade citizens and others from speaking out, being active or responding in kind and employing the intense forms of lawfare and imprisonment that the Democrats employ to further their goals.

A less obvious abuse of our legal system is the use of lawfare and administrative process to cover up election fraud or other legal abuses within the bureaucracy; to attack those who seek relief from those abuses or to subvert legitimate reform or election security efforts. In this form of lawfare, public officials use ostensibly "lawful" acts such as threats of prosecutions, impeachments and administrative attacks to aid and abet illegal acts and crimes that may have already occurred or to create or maintain insecure election processes.

Examples of this abuse include the prosecution of Tina Peters and the impeachment of Ken Paxton Attorney General of Texas. Another example of this form of an abuse of prosecutorial authority is the effort by the Federal Department of Justice to shut down the investigation that was conducted by the Arizona

Legislature regarding the 2020 election both before and after January 6th, 2021.

Another example of this form of lawfare are Federal lawsuits initiated by the Department of Justice against States that attempt to require voter identification or purge their voter rolls of non-citizens and other ineligible persons.

When the Arizona legislature conducted its own investigation, which was clearly its Constitutional duty, it began a canvassing effort. This canvassing effort involved going out into the field and verifying the addresses of alleged "voters". It was quite obvious from looking at the voter rolls that there were large numbers of so called "voters" whose "residences" were vacant lots, commercial buildings, post office boxes or multiple voters at the same address. The purpose of creating fake or ghost voters is to generate a blank mail in ballot that can then be filed in by criminals seeking to change the outcome of the election. It does not take a large number of people engaged in criminal activity to change the outcome of the election. These ghost "voters" clearly existed but could it be documented that their numbers were significant enough to change the outcome of the election? A canvass threatened to document with statistically verifiable and reliable data that the numbers of "ghost voters" was indeed significant and easily enough to change the outcome of the election. Unofficial canvassing efforts had, by December of 2020, already indicated that there were significant numbers of "ghost voters" in Arizona. There were also numerous other violations of election law. If the Arizona legislature itself issued a report that included canvassing data the report would be powerful evidence that the number of illegally cast ballots was significant and enough to change the outcome of the election. This would be an extremely dangerous development because it could lead to the Arizona legislature decertifying the results that had been previously certified by the partisan Secretary of State but had never been certified by the State Legislature.

This was indeed a crisis so the Federal Department of Justice had to act quickly to arrest this investigation. In order to stop the investigation, the Department of Justice had to intimidate members of the Arizona legislature and threaten them with Criminal Prosecution if they persisted in using canvassing to assess the validity of the certification that Arizona had submitted to the Congress. These "lawful" administrative actions may have been designed to protect the potential criminal acts that may have already occurred. The Justice Department may have literally been aiding and abetting criminal vote fraud by preventing effective legislative investigations.

These actions may have been a deliberate weaponization of Federal law enforcement designed to aid and abet criminal vote trafficking by preventing the discovery of vote crime. It is similar to hiding a murder weapon by someone who themselves may not have participated in the initial murder. Any actions by state or Federal actors designed to protect vote fraud in Arizona are potentially criminal acts of aiding and abetting as an accessory after the fact. See, Arizona law Rev. Stat. sec. 13-303 and potentially a crime under Arizona law Rev. Stat. sec. 13-306.

The Impeachment of the Attorney General of Texas

The impeachment of the attorney General of Texas Ken Paxton, a heroic freedom fighter, is an indication of just how serious the crisis is that the American people currently face. The impeachment of Ken Paxton is also an example of how dangerous the behind the scenes alliance between Democrats and rino republicans truly is. Attorney General Paxton had vigorously fought the lawfare to dismantle ballot security measures in Texas prior to the 2020 election. Paxton was also effectively using the judicial system to impede the Biden Harris administration's efforts to weaponize migration, open the borders and flood the United States and Texas with economic refugees, criminals, terrorists and military forces from hostile powers. Paxton had to be attacked and removed in part because he was so effective at supporting free and fair elections and opposing government corruption and the invasion of the United States.

In early 2021, Paxton's office received complaints from FBI insiders about illegal activity within the woke FBI. Attorney General Paxton began an investigation which was his duty. The FBI knew about his investigation before it even started and immediately began efforts to sabotage and undermine the investigation. Paxton's staff was pressured into resisting his investigation. Paxton was told they were not going to investigate because law enforcement can't commit crimes. So, Paxton appointed a special prosecutor.

In order to stop Paxton, the Democrats had to bring an impeachment proceeding in the Texas legislature. However, in order to impeach Paxton, they needed the support of some elements within the Republican party. The Democrats have significantly controlled the Texas legislature for years by exploiting divisions within the Republican party and making deals with misguided and selfish Republicans who sell out their party and country in return for some meaningless political concession. The concession that the Democrats demanded to install their favored Republican Speaker of the Texas house was for the Republican speaker to join in and facilitate the abuse of the impeachment process against Attorney General Paxton. This impeachment effort was a clear and direct attack on popular sovereignty. This attack was launched because Paxton was serious about using law enforcement to secure free and fair elections and fight vote fraud and corruption within the ranks of law enforcement. He was also very effective at holding the Biden administration accountable for its violations of Federal Immigration law. He had to be stopped and an example made of anyone in a position of power who dared to insist on election integrity. That is the true reason Paxton was attacked.

Attorney General Paxton actually was impeached but was not convicted by the Texas State Senate and therefore remained in office. The sad and even more disturbing aspect of this assault on popular sovereignty and election integrity was that many Republican neo cons joined the effort. It appears that the Bush apparatus in Texas was cooperating in and aiding and abetting the attack on Ken Paxton. This collaboration is extremely concerning because it is evidence that there are elements within the establishment Republican party that are willing to make common cause with and empower a Democrat party that is dedicated to transforming the American Republic. That is the essence of the problem American citizens face not only in Texas but nationwide as well.

When the Cheneys both Liz and Dick endorsed Kamala Harris it brought out into the light and illustrated this truth that many have suspected for years. The current Democrat party is too dangerous for Robert F. Kennedy Jr. and Tulsi Gabbard but it is not too dangerous for the Cheneys. That is how powerful the globalists are. In fairness to the Cheneys and other rinos this endorsement came about a week or 2 before Biden's controllers decided that it would be a good idea for the United States to launch sophisticated cruise missiles directly into Russia without Congressional authorization.

This lawfare is a serious matter in part because it is not initiated in good faith. The contention that the prosecutors are acting in the public interest pursuant to legitimate public authority is a lie. If the deep state succeeds in silencing its opponents through intimidation and lawfare there won't be any effective barrier to prevent them from ending our Constitutional Republic and forcing America into a post Constitutional era.

The Democrat party makes no secret of its desire to politicize the judiciary starting with packing the Supreme Court and creating an oversight bureaucracy to intimidate Supreme Court Justices. The Supreme Court is the one remaining vestige of Constitutional authority that has not been intimidated and overtly politicized. The politicization and destruction of the Supreme Court would be a very dangerous development because it would mean the removal of the one remaining Constitutional barrier to the destruction of our Constitutional Republic.

The Biden regime has also used both the Justice department and intelligence agencies to attack free speech. Early in the Biden administration there were official proposals to create a censorship bureaucracy to regulate "misinformation". Those proposals have currently been put on the back burner for strategic rather than philosophical reasons. No doubt the proposals will be renewed if they succeed in eliminating the Supreme Court. In the spring of 2024 we learned that the FBI and the CIA had both been involved orchestrating big tech efforts to censor information the regime did not want publicized. Government agencies had also been involved in the very beginning in supporting the civil lawfare that was orchestrated by the Democrat party in an effort to bankrupt and silence Alex Jones and Inforwars and remove Alex Jones from the public square. There have been ceaseless and ongoing lawfare efforts to bankrupt and shut down other conservative media outlets as well such as the Epoch Times that refuse to support the regimes narrative. We will learn much more about this abuse of public trust and authority in the future as the investigations and civil RICO suits against the U.S. government proceed.

The lawfare against Donald Trump and his associates and allies is clearly an organized and willful abuse of the prosecutorial function. These attacks may also be a violation of the RICO laws, Racketeering and Corrupt Organizations act. Time will tell but there should be a serious RICO investigation and known enablers subpoenaed to testify before Congress and in DOJ inquiries. These outrageous abuses of the prosecutorial process are not acceptable and probably not lawful.

Our system of justice is precious and not a political tool to play around with and abuse. The true purpose of the onslaught of lawfare against the President, from 2 impeachments to baseless criminal charges to civil forfeiture proceedings to finally a plain, open effort to remove the President from the ballot in several critical battleground states, is to attack our democratic process. None of this lawfare is initiated in good faith. We now know that much of this lawfare was orchestrated and coordinated through the Biden White House and Department of Justice.

This weaponization of government, judicial and prosecutorial functions for partisan political ends is totally unacceptable in a free and functional Constitutional Republic. This abuse of justice is a well-coordinated and broad-based attack on both the American Republic and on all of us using lawfare and criminal prosecutions of the leaders of the resistance as a weapon to silence and intimidate anyone who dares to demand free, fair and auditable elections or to object to the destruction of the American Republic. This attack on the American Republic effects all of us including the progressive liberals, flag burners, antifa, BLM and others who naively assume that if we destroy our Republic, our Constitution and system of justice there won't be any adverse consequences for them.

The weaponization of law enforcement and the administrative state by a government that is not acting in good faith and may not even legitimately possess the consent of the governed is a form of soft violence against the American Republic. The theft of the 2020 election is violence against the American Republic. The use of extreme lawfare against President Trump to deprive the American people of their duly elected President and to prevent his reelection is a willful attack against the American Republic and the American people. The theft of the 2020 election has already done extensive and perhaps irreparable damage to the American Republic. The opening of our borders and turning control of the borders over to criminal drug and human traffickers is not merely an impeachable dereliction of duty it is an act of violence against the American Republic which will result in chaos, violence and ultimately may contribute to or cause the collapse of the American Republic.

The weaponization of justice is just one part of a process designed ultimately to eliminate citizen sovereignty and to prevent American citizens from regaining control of their Republic and their future. The critical point right now for the American people to understand is that these abuses are not the normal operation of an unbiased system of Justice. The multiple attacks on President Trump are not accidents or coincidences. These attacks are deliberate, willful and purposeful. These attacks are attacks on all of us. This attack on the Republic is totally unacceptable. These attacks cannot go unanswered. They must be fought vigorously. The perpetrators of these attacks must be held to account.

CHAPTER SIX

THE FEDSURRECTION

There has not to date been a serious impartial investigation of January 6th 2021 and the events of that day. The Democrat committee that purported to conduct an investigation destroyed some of the evidence. It appears the FBI may also have damaged or defaced evidence of the surveillance video from the DNC headquarters showing the planting of the fake bomb. As more and more evidence comes to light it is quite clear there were numerous federal agents or operatives that were present. The exact number has not been publicly available yet. We know there were at least 8. Many commentators have estimated that there were 50 or more federal agents or operatives at the Capitol grounds on January 6th. Some commentators have also speculated that part of the reason all of the security camera video from January 6th has not been released is because it could lead to identification and a more accurate count of federal operatives present on Jan 6th.

Of the videos that have been released so far, agents or operatives dressed as Trump supporters can be seen being "arrested" handcuffed, lead inside the Capitol building and then released by Capitol Police. The Q shaman can be seen being escorted by Capitol police into the house chambers. Capitol police can be seen beating demonstrators including Roseanne Boyland who was beaten to death by Capitol police.

As January 6th 2021 began the President had not yet been informed that Vice President Pence was not going to support the Presidents teams' plan to present evidence of the massive irregularities and even fraud that had occurred in the 6 key battleground States of Arizona, Pennsylvania, Georgia, Wisconsin, Nevada, and Michigan and to request that those reported results not be certified by the Congress on the grounds that the ballots used to support those certifications cannot be verified to be lawful votes. There is nothing in the Constitution that requires Congress to ratify fraudulent election results. The President's team could potentially have forced a vote as to the matter of whether the certification by the State should be returned to those States for appropriate further legislative action. Yes, it would have been better if those State legislatures had acted before January 6th however, unfortunately, that did not happen. The end result of this strategy could have been that the House of

Representatives would decide the election if no candidate reached 270 electoral votes. It could have been the electoral vote would be postponed until as late as January 19 or even later due to the exigent circumstances that had been presented by COVID.

The Presidents team was to be allotted 2 hours per State to present evidence as to why suspect certifications should not be accepted or approved by the Congress and instead be returned to that State for further consideration by the State legislature. There would have been 12 hours to present evidence. The last thing that the Presidents team wanted was a disruption of the Congressional certification of the election. It was the only path available to use lawful due process to enforce the verdict of the citizen electorate. This was the logical and lawful response to the extreme lawfare initiated by the Democrats. Even if the likely end result would be a vote to accept the certifications by the partisan Secretaries of State, in a solemn proceeding in Congress, the media would not be able to ignore or suppress the evidence of uncertifiability and fraud as it had since Nov. 3, 2020. 4 State legislatures had pending proposals for legislative action to correct the certifications of the 2020 federal election that their partisan Secretaries of State had issued. Unfortunately, the fedsurrection began literally at the moment Congress was to begin hearing evidence. The riot essentially ended the process of informing the American people about what had happened in the 2020 election.

The fake attempted kidnapping of the Governor of Michigan Gretchen Whitmer is a clear indication that the FBI and other deep state elements are quite capable of and willing to cultivate false flag operations to propagate the ridiculous lie that white supremacists are somehow a threat to the American Republic. The FBI cultivated and grew the Whitmer plot and then stopped it before it was to be executed. Of the people, 28 or so people involved in the Fednapping caper several of them, including some in leadership positions, were agents or people bribed by agents to participate in the plot.

There were numerous actors on Jan. 6th who were not in any way associated with Trump. These groups included at last some federal agents or operatives dressed as Trump supporters and Antifa elements impersonating Trump supporters. There were 3 main "right-wing" groups associated with planning a disruption on January 6th. These groups were: The Proud Boys an independent right-wing group not associated with Trump or the Maga movement; the Oath Keepers, a more religious-oriented group also not associated with Trump, and the Three Percenters.

The FBI or other federal agencies had infiltrated all of these groups before the Whitmer plot and well before January 6th.

Revolver News has done extensive reporting on January 6th and other related issues. In one of its articles on this topic, it points out the implausibility of the theory that January 6th was caused in part by intelligence failures and therefore the FBI and other intelligence agencies could not prevent the disruption.[10] The same 3 groups present on January 6th had also been involved in the Whitmer plot and in that instance, the FBI had infiltrated all 3 groups. Following the rollup of the Whitmer caper the lead FBI special agent in charge was then transferred from Michigan to Washington DC before January 6th.

> *Let's recap what we've established. Just months prior to the U.S. Capitol Siege on 1/6, the FBI thwarted a similar plot involving a siege at the Michigan State Capitol, whose plotters belong to one of the three main militia groups associated with 1/6. The FBI was able to thwart this on the basis of an astonishing infiltration rate of said groups involving undercover operatives and informants who had been working in such capacity, just in one tiny Michigan network, for more than seven months. They were so well-infiltrated that they already had three informants embedded in this random Three Percenter network before*

---

[10] Unindicted Co-Conspirators in 1/6 Cases Raise Disturbing Questions of Federal Foreknowledge - Revolver News (June 14, 2021).

any plot was even hatched. Furthermore, just days after the plot was foiled, FBI director Christopher Wray quietly promoted the FBI Special Agent in Charge of the Michigan Plot operation to a coveted D.C. field post, where he now oversees the investigation into 1/6.

The Special Agent in Charge, by the way, is who establishes, extends, renews and supervises all FBI undercover operations.

The above parallels between the Michigan Plot and 1/6 do not necessarily mean that the FBI had undercover informants and operatives who were involved in 1/6. But it sure as heck reinforces our intuition that it's a distinct possibility. And it forces us to ask the question once again — if the government foiled the Michigan Plot, why didn't they step in to stop the so-called siege on 1/6?

It is now imperative for anyone who cares about the truth to demand that Christopher Wray answer the question — to what extent did the FBI or any other government agency infiltrate the key militia groups associated with the U.S. Capitol Siege?

And more pressing still, a question to which we now turn our attention: how many of the unindicted co-conspirators in 1/6 prosecutions are unindicted on account of a prior arrangement with the federal government as an undercover operative or informant?

At about 10:50 in the morning of January 6th, 2021 an undercover Capitol police officer discovered a pipe bomb adjacent to Democratic party headquarters. The "bomb" appears to have been sitting there for several hours. The bomb was reported to Capitol police and Secret Service officers who were on-site at the time. There was also another bomb found at RNC headquarters. The timing devices on the "bombs" were set for 1:00 PM the time at which Congress was to convene to consider certification of the election. Many police were relocated from the Capitol to party headquarters. Subsequent investigations through Representative James Comer's oversight committee in the House by early 2024 had concluded that one of the fake bombs was discovered by a Capitol police officer. The bombs were fake and it appears that the Secret Service on the scene knew or concluded that the bomb at the DNC was fake before the device was removed by a robot. The original video of the person planting the bomb has not been released. That person's identity is currently unknown. It appears that the FBI has not to date made a serious effort to determine the identity of the person. Was the fake bomb also planted by federal agents or operatives?

Vice President elect Kamala Harris was in the DNC headquarters at the time. There was no need to evacuate her, clear the area, or take any action to protect the kids that were allowed to roam about near the site where the bomb was found. It appears that both the Secret Service officers detailed for the security of Vice President Kamala Harris and the Capitol police were aware the bomb was a fake. No bomb squads were called in to investigate until later. The area was not evacuated. However, Capitol police were redeployed from the Capitol to the DNC.

The President gave a speech and addressed a very large crowd at the ellipse near the Capitol building. The President began speaking at the ellipse at about 11:40 AM on January 6th, 2021. It was quite clear that the President was urging the crowd to "peacefully and patriotically" demonstrate their support of the members of Congress who were moving to not accept the certifications from States where the evidence of unverifiability and fraud was so significant that the results were unreliable and therefore should not be included in any official action that needed to be founded on verifiable, authentic votes. One key candidate for decertification was Pennsylvania where a partisan Secretary of State had certified an election result that included numerous violations of Pennsylvania election law and there were proposals that the Legislature consider its own certification.

While the President was still speaking police removed the bicycle racks that had been placed around the Capitol steps on the west side. At approximately 1:05, just as Congress was set to convene, on one side of the Capitol building, a very small number of the crowd began to enter the Capitol grounds without any police opposition. These people included Antifa elements dressed as Trump supporters.

We know there was at least some involvement by federal agents or operatives in the riot that occurred on January 6th, 2021. The American people likely will not know the extent of federal involvement on January 6th until there is a serious investigation. We know that there were some people dressed as Trump supporters who were associated with other organizations including radical groups. We know that the Speaker of the House, who was in charge of security, refused President Trumps recommendation that 10,000 National Guard be deployed on January 6th, 2021.

There is a report by the Inspector General's office that reportedly contains information about the involvement of Federal operatives on January 6th, 2021. The report has not been publicly released and probably will not be released until after the 2024 election.

The investigation process has begun with house committees such as Representative Comer's Oversight Committee but much more needs to be done. It appears that federal agents or operatives may have been involved in or even were responsible for constructing the gallows that were supposed to be used to hang the Vice President. Up until now, when asked directly how many federal assets or agents participated in January 6th, deep state bureaucrats have been allowed to simply refuse to answer without any legitimate reason and then go on their merry way continuing to draw their massive salaries. When patriots like Steve Bannon or Peter Navarro refuse to answer questions—with legitimate reasons— they are given prison sentences for contempt of Congress.

By September of 2024, Congressional committees had determined there were at least 8 Federal agents or operatives present or involved in January 6th.

There were also some 50 or more operatives some dressed as Trump supporters that arrived in 3 buses. More investigation needs to be done of those people and what organizations they were associated with. Another suspected operative who had publicly, repeatedly urged the crowd to enter the Capitol, spoke to a young man in Trump garb who began removing one of the bicycle racks that had been placed as a barrier by Capitol police. This person apparently encouraged the first movement by the crowd to cross police barriers. This person was initially on FBI wanted websites then quietly removed. He was eventually given a slap on the wrist and allowed to plea to misdemeanors when the extent of his involvement became widely known and he had to be charged with something. He had to be charged with something in order to maintain an appearance of impartiality and prosecutorial integrity. The Department of Justice could not continue to protect this person and allow him to walk free while subjecting dozens of other January 6th defendants to solitary confinement, beatings, torture and lengthy prison sentences.

On July 13, 2024 there was the first attempted assassination of President Trump in Butler, Pennsylvania. There is already significant evidence to indicate that there must have been some assistance to the shooter and it may have come from rouge elements within the deep state, FBI, CIA or other forces. The investigations are just starting but there are already many serious questions and inconsistencies. We know there are at least some elements within the deep state willing to engage in criminal conduct and false flag operations. What basis is there to believe that these elements would somehow shy away from orchestrating a riot on January 6th, 2021.

CHAPTER SEVEN

UKRAINE CONFLICT

The situation in the Ukraine could easily escalate into WWIII if this war isn't resolved soon. Some say WWIII has already started. The American people have been given very little information to understand this complicated and explosive situation. How did we get here? Is it possible that the deep state could involve us in WWIII without authorization from Congress. The Ukrainian situation probably could have been resolved before now. The situation is rapidly getting out of control. One piece of information we do have come in the Trump/Harris debate. Trump alleged that Biden had not talked to the Russian leadership in over 2 years. Harris did not deny the allegation and then went on to explain that war with the Russian Federation might be necessary despite the incredibly weak position America is now in thanks in part to her and Joe Biden.

The ties between Russia and the Ukraine are very deep and longstanding. The Ukraine has always historically been a part of Russia since the 1500's and later the Soviet Union until 1989. The Ukraine is the border area between Russia and Europe.

The Ukraine was not an independent country until 1991. It is fair to say that Ukrainian and Russian relations have been strained and hostile at times. They have been fighting over these borders and lands for a long time. At other times they have worked together sharing a common culture, language, region and religion of Orthodox Christianity.

Imperial Russia was an Orthodox Christian Empire. One of the issues that caused WWI and was a significant issue in the aftermath of both WWI and WWII was the aspirations of autonomy and independence of national groups that had formerly been a part of the Austro-Hungarian, Russian, French, Ottoman or British Empires. The current situation with the Ukraine is evidence that these aspirations of national independence still cause tensions within the international community to this day.

In 1989 the collapse of the old Soviet Union created a new opportunity for Ukrainian independence. The current situation in the Ukraine should be seen both as a conflict that has developed over hundreds of years and a conflict that is inexorably tied to the dissolution of the old Soviet Union. According to Russia, the Ukraine itself is a creation of the Soviet Union that was always a part of greater Russia itself not just a part of the former Russian empire. Events in Ukraine inevitably involve Russia's national security interests. It is reasonable that Russia be concerned with what happens in the Ukraine.

There have already been 3 invasions of Russia in the modern era, 1812, 1915 and 1941. Russia has repeatedly warned us since 1989 both publicly and privately that the Ukraine cannot join NATO as that would seriously compromise Russia's vital national security interests. Russia could not allow NATO forces to be stationed on either its border with the Ukraine or in the Crimea. Such a move by NATO would be the beginning of WWIII. That is how serious the matter is to them. We have continually resisted taking Russia's fairly reasonable security concerns and warnings seriously.

The Biden administration's refusal to negotiate in good faith with the Russian Federation about its security concerns has resulted not only in a serious and unnecessary risk of nuclear war it has also resulted in Russia abandoning the West, the U.S. dollar and embracing closer ties with the Chinese communist party, Iran and North Korea. This development is a strategic and economic catastrophe that should have been avoided at almost all costs.

To appreciate the enormity of the strategic catastrophe that has occurred over the last 35 years and culminated under Joe Biden, we should look at American and Russian relations since the end of WWII. By the mid 1950's with 3 major nuclear armed powers in the world, the U.S., USSR and China, it was understood there was a serious danger that any future great power conflict would almost certainly involve the use of nuclear weapons. This strategic background was present throughout the cold war. Another aspect of this strategic background was that if 2 of the 3 great powers acted together or in concert the odd man out would be at a serious and probably insurmountable strategic disadvantage in the event of war or violent confrontation.

Russia and the United States have much in common culturally. Many of the people share a Christian faith. The people share many common cultural interests of family values, the arts, sports and industriousness. We share a common interest with the Russian people in defeating Muslim supremacy and preventing the Islamist goal of worldwide Islamic rule. Russia understands much better than we do how to deal with Muslim extremists. The vast majority of the American people also share a common interest with the vast majority of the Russian people in defeating progressive globalism and securing an international order that respects and supports sovereign nation states as the building blocks of a peaceful international order.

It is essential that sovereign nation states survive even if they aren't purely democratic in order to resist progressive globalism and transhumanism. If the sovereign nation states throughout the world are not preserved and militarily secure, sovereign states may ultimately be faced with an impossible choice: surrender sovereignty to a new world order or face conquest and repopulation.

For years our leaders have said that NATO was not an offensive organization. NATO was a defensive organization. And yet NATO attacked the Serbs in Kosovo and continually expanded Eastward.

Cooperation and focusing on shared interests and values could have yielded significant benefits for both sides. There were proposals for the Russian Federation itself to join NATO. This might have been possible if there had been a comprehensive security and trade agreement between Europe, the EU, NATO and

Russia. This goal may have been difficult to achieve but it was certainly one worth discussing. A comprehensive security agreement with the Russian Federation could have strengthened ties within the Judeo-Christian world and put the West in a far stronger position vis a vis China and Iran than we are currently in. The true cost of our misguided and unnecessary policy to expand NATO eastward to Russia's border has been to lose the possibility of a much stronger and more positive relationship with Russia and, worse yet, to isolate the United States.

The complexity and difficulty of Ukrainian/Russian relations is exemplified in some of the negotiations regarding the treaty on the eastern front at the end of WWI. Most of the area of the Ukraine was separated from Imperial Russia by the treaty of Brest-Litovsk which ended Russia's participation militarily in WWI. The Germans gave Lenin gold and allowed him to cross German lines because they knew the Bolsheviks would undermine the Russian military effort if not cause its collapse. Bolshevik propaganda had endorsed an end to the war by any means necessary. The Eastern front did collapse and Germany was able to send its forces further into Russia without any significant military opposition. The Bolsheviks wanted a ceasefire on all fronts but that was not realistically possible at that time.

Germany wanted to send its forces on the Eastern front to the West in anticipation of imminent U.S. involvement in the war.

There was an armistice on the Eastern front in December of 1917, 11 months before the general armistice in WWI in November of 1918. In February of 1918, Germany and the Central powers concluded a treaty with The Ukrainian Peoples Republic which was headquartered in Kiev. This first treaty at Brest-Litovsk provided for Ukrainian independence from Russia. The Russian/Bolshevik regime did not recognize the Ukrainian Peoples Republic. The Bolshevik favored regime in the Ukraine was the Ukrainian Peoples Republic of Soviets in Karkhov. The Bolsheviks refused initially to agree to the treaty that had been negotiated by the Ukrainian Peoples Republic. The Germans then renewed operations on the Eastern front. Within a few weeks, an agreement was reached with the Bolshevik regime in March of 1918. In return for the Bolshevik approval of what was essentially the first treaty negotiated by the Kiev regime, Germany withdrew the bulk of its forces and did not proceed further into Russia following the collapse of the Eastern front. Obviously, the Bolshevik regime did not want to surrender the Ukraine and did so only under extreme duress.

The Ukraine enjoyed some degree of autonomy or at least was not under Bolshevik control for a few years after the War. In November of 1917, the Bolshevik faction closed down the Constitutional assembly that had been elected to draft a constitution for Russia. This attack on the Constitutional assembly was associated with the beginning of the Russian civil war which lasted in some parts until almost 1925. The Ukrainian Peoples Republic emerged almost immediately in 1917 in an effort to secure Ukrainian independence from Russia which at that time was not yet under Bolsehvik control. There were different Ukrainian factions including White Russian elements who maintained a military opposition to the Red Army. The Soviet-Ukrainian War lasted from 1917 to 1921. There were Ukrainian forces on both sides of that war and also in the broader Russian Civil War. The Red Army defeated the various opposition and Ukrainian factions in 1921. The Ukrainian Socialist Republic joined the Soviet Union in 1922 as one of the founding socialist republics of the Union of Soviet Socialist Republics, USSR.

The treaty of Versailles finalized the terms of the armistice which ended WWI and was negotiated over a period of several months ending in early 1920. The Bolsheviks choose not to participate in Versailles in part because they believed a general world revolution was imminent. The Versailles powers however did protect Russian imperial interests including Russian sovereignty in the Ukraine. The Great powers appear to have believed that the White Russian forces would eventually be victorious in the Civil war. The issue of Ukrainian independence could then be taken up after the Bolsheviks were defeated. Unfortunately, the Bolsheviks were eventually victorious. The Versailles treaty in essence reversed the treaty of Brest Litovsk with respect to Ukrainian independence and allowed for continued Russian hegemony in the Ukraine rather than supporting Ukrainian independence.

The Ukraine played a very important role in financing the new Bolshevik regime. Russia was not an industrial society at that point and needed hard currency in order to finance industrialization and the continued existence of the Soviet regime. Significant amounts of hard currency were secured by confiscating and selling Ukrainian and also Russian grain in the world market. Unfortunately, this policy resulted in massive famine in the Soviet Union and Ukraine. Millions of people starved to death in the Ukraine and Russia in the early 1930's. It has been estimated that as many as 7 million people died of starvation and malnutrition just in the Ukraine in the 1930 to 1933 time frame.

When Germany betrayed the German-Russian non-aggression treaty and secret protocols of 1938 and invaded Russia in June of 1941, the Ukraine was the first area of the Soviet Union to become a battlefield. It is true that some, but certainly not all, Ukrainian people welcomed the German army as liberators. Some Ukrainians fought with the German army believing that Germany would liberate the Ukraine from Russian rule. Many Ukrainians also fought with the Russians. Ukrainian elements became an important 5th collum of the Red Army behind German lines in the Ukraine. It is fair to say that most but not all Ukrainian people fought for and supported Russia's struggle to defend the USSR from German aggression.

In the end however, Russia did not give up the Ukraine in WWII despite extreme duress and violence from the German military. The German army was defeated in Russia by the Red Army and the people of the Soviet Union including most Ukrainian people.

Before the final end of the War, the German army was expelled from the Soviet Union which at that time of course included the Ukraine. Fighting between Soviet and Ukrainian forces began in the background and continued in various forms for several years. The main Ukrainian opposition forces were the military forces who had fought with the Germans. Some of these Ukrainian groups were neo Nazi forces associated with Stephan Bandera who is a hero to some contemporary Ukrainian nationalists. There were also other Ukrainian nationalist forces and most rejected neo Nazism especially after WWII ended. Many nationalist opposition elements were sent to prison camps in Siberia and elsewhere. Fighting in the Ukraine continued well into 1950s.

1989 Dissolution of the Soviet Union

In 1989 German unification began. According to multiple sources one of the significant behind the scenes developments that made German unification possible was the willingness of the Soviet Union to withdraw to back away and allow East German independence less than 45 years after the German invasion that sought to achieve a mass extermination of the Russian slavic people. This withdrawal was tied to an understanding that there would not be an Eastward expansion of NATO. This understanding was in many ways a win situation. Had we been willing to respect this understanding there is no telling where the world could be now. It is possible that an area wide security agreement involving both NATO, the EU and the old Soviet Union could have been achieved especially after 1989.

In 1989 the Soviet Union dissolved as it became evident that a significant portion of the Russian military was unwilling to be used against the Russian people. When the dust settled, many parts of the former Soviet Union formed the Russian Federation.

In 1991 the former Ukrainian Socialist Republic became a quasi-independent country. The Ukraine's original Constitution called for it to remain neutral. For the first few years of Ukrainian independence the Ukraine retained a significant number of nuclear weapons that had formerly been part of the Soviet Union's nuclear capabilities. The neutrality of a former Republic of the Soviet Union that possessed Soviet nuclear weapons was certainly a significant issue to Russia. Russia could never allow a situation in which Soviet nuclear weapons were turned over to NATO on Russia's doorstep. This reason is partly why Russia has been consistent in insisting that the Ukraine cannot join NATO. NATO ICBM's and medium range missiles could easily reach Russian territory from the Ukraine and NATO anti-ballistic missile batteries would be in a position to shoot down Russian ICBM's in their launch phase. Deployment of these military assets would give NATO and the West a first strike capability and would radically upset the strategic balance of power.

Budapest Memorandum

There were also other ongoing issues in addition to the nuclear weapons in the Ukraine. During the Clinton Administration, from 1992 to 1994, there were some agreements crafted known as the Budapest memorandum. This was not a formal treaty but was a formal understanding in which the United States, United Kingdom and Russia guaranteed Ukraine's sovereignty. The United States also agreed to give "assurances" of Ukraine's security and territorial integrity if Ukraine gave up its nuclear weapons. Obviously, President Clinton did not have the authority to unilaterally bind the United States to such a commitment. However, Ukraine did agree to decommission its inherited Soviet-era nuclear weapons probably to some extent due to assurances from the United States to support Ukraine's security. This was a kind of back door way of providing for Ukraine's security without formally admitting the Ukraine to NATO.

Origins of the Current conflict

In the period of time between Ukrainian independence in 1991 and the outbreak of larger scale violence in the Donbas region in 2011, Ukrainian oligarch and organized crime factions were able to gain influence throughout the Donbas region and the Ukraine. The prevalence of criminal and neo Nazi elements of extreme Ukrainian nationalism may have been one of the reasons why many people in those disputed areas wanted a status that was independent from the Ukraine.

The current phase of violent conflict in the Ukraine started in 2014 in the Donbas region of what was then part of the Ukraine. The Donbas region is an industrial area in far Eastern Ukraine. The disputed regions that have been fought over since 2014 are part of the Donbas area. These areas are; Donetsk and Lugansk which had been a subject of the first Minsk accords. Other disputed regions included Zaporizhzhia and Kherson. Russia had already secured the Crimea in 2014 in the aftermath of the Maidon coup which deposed President Yanukovych.

In 2014 the pro-Russian elected president of the Ukraine, Victor Yanukovych was deposed and replaced with a Ukrainian nationalist. This was the so called Maidon revolution. It began as a civil disruption when Yanukovych refused to approve Ukrainian membership in the European Union. Ukrainian membership in the EU was a somewhat complicated issue and one of the reasons the Minsk accords were negotiated. There are numerous trade and national security issues which are implicated in Russian/Ukrainian relations.

When President Yanukovych did not approve the EU membership it touched off civil disturbances in February of 2014. These disturbances escalated to confrontations with police. There were Banderist and other extreme nationalist elements who were involved in the overthrew of an elected President of the Ukraine. According to western and U.S. versions, Yanukovych resigned. Another explanation of Yanukovych leaving office was that he had to flee Kiev and go to the Donbas region in order to avoid being arrested or killed.

It seems fairly clear that the U.S. was involved in the overthrow of Yanukovych. U.S. involvement has apparently never been officially denied. Russia likely interpreted American involvement in the overthrow of Yanukovych as a violation of the understanding of the Budapest memorandum in which the U.S. agreed to guarantee Ukraine's sovereignty and also was another in a string of violations of the post-cold war understanding from 1989 that there would not be eastward expansion of NATO. By our actions we sent a clear indication that the West was continuing to push NATO eastward. Our support of extreme Ukrainian nationalism could easily and reasonably have been interpreted by Russia as preparatory to continued NATO expansion Eastward with an obvious goal of including Ukraine in NATO which we knew was a major red line for Russia. It was certainly plausible that other acts of aggression against Russia including possible "regime change" involving the Putin government were at least contemplated if not actively pursued.

Fighting had been going on in Donbas region between Russian para military separatists and the Ukrainian para military and regular military since at least 2011. Russian separatists objected to what they viewed as oppressive conditions imposed on Russians in the Donbas. According to the Russians, the Ukraine tried to prohibit the Russian language and even pursued ethnic cleansing in those areas. Prior to the Russian military being sent into the Donbas region in February of 2014, some 14,000 or more people were killed in the first phase of the war in the Donbas. Many Russian people in the Donbas wanted the region to be reabsorbed into the Russian Federation or at least be administered by the OCSE (Organization for Cooperation and Security in Europe) pursuant to the terms of the Minsk accords.

Following the overthrow of President Yanukovych, almost immediately, Russian separatists declared independence from the Ukraine and set up the Donetsk Peoples Republic (DPR) and

Lugansk Peoples Republic (LPR). Regular Russian military was sent into the Donbas and Crimea in February of 2014. Russia did not recognize the DPR or the LPR or incorporate them into the Russian Federation at that time.

Each side, Russia and the Ukraine and other signatories to the Budapest Memorandum accused the other of violating the agreement in the aftermath of the Maidon coup in 2014. Russia was accused of violating the territorial integrity and sovereignty of the Ukraine by invading the Crimea and Donbas regions in 2014. According to Russia, the Ukrainian State that executed the Budapest Memorandum no longer existed. The West had violated the agreement in several respects including through economic coercion. According to Russia, the overthrow of President Yanukovych was more than just a violation of the agreement. The overthrow had ended the Ukrainian State. Moreover, Russia was not obligated under the agreement to participate in forcing parties to be subject to the Ukrainian state against their will in reference to the DPR and LPR.

The issue of sovereignty in the Crimea was itself quite complicated. According to Russia's position, the Crimea was not even considered part of the Ukrainian Socialist Republic until the mid 1950's and even that was not settled within the Soviet Union.

Later as the Soviet Union was breaking up, in order to obtain International Monetary Fund funding, Soviet era officials executed agreements that purported to renounce territorial claims to the Crimea. According to Russian nationalists, Soviet era officials did not have the authority to execute such an agreement because the Soviet Union was breaking up and their authority derived from the old Soviet Constitution.

Russia claimed it had to militarize the Crimea to preempt any possible move by the Ukraine to allow NATO forces into the Crimea itself. There were extensive Soviet era naval bases in Crimea that Russia could not risk falling into NATO hands. The Crimea had been part of old Russia and then the Soviet Union until 1989. The Crimea was Russia's main warm water port in Western Russia. Russia had fought a war in the 19th Century in part to secure the Crimea from British, French and Turkish occupation. In an environment in which all parties were acting reasonably and in good faith it is likely that Russian sovereignty in Crimea would be recognized as part of an overall security arrangement. This militarization of the Crimea happened while Biden was vice President and there was barely a peep out of the Obama/Biden Administration.

There was ongoing violence in the Donbas region from 2014 until 2022 when there was a larger scale Russian military intervention. In September 2014 the first Minsk Accord between the Ukraine, Russia the DPR and the LPR created a ceasefire. This first ceasefire was short-lived. It seems clear that Russia was supplying separatists in the DPR and LPR with soldiers and military equipment throughout this period. The first and second Minsk accords called for a plebiscite in the 2 disputed regions and for those regions to be quasi-independent regions ultimately under Ukrainian sovereignty with the cooperation and oversight of the OCSE. The Minsk accords were also an effort to reach agreement on a number of issues such as trade. Subsequent election results in those 4 areas indicated that the people supported the Minsk accords and wanted at a minimum a quasi-independent status. Although there may have been some controversy about the results, these were Russian speaking areas that had always been part of greater Russia or the Soviet Union until 1991.

One of the problems with the Minsk accords was that the U.K. and U.S. along with Ukrainian Nationalists did not want the Minsk accords because the Minsk accords would ultimately involve Ukrainian neutrality. Neutrality would mean the Ukraine could trade with either East or West, Russia or Europe and ultimately not be part of NATO. The accords also provided for the removal of heavy armaments from the area. Russia also dropped objections that the Ukraine could join the EU which was one of the issues that had touched off the coup in 2014.

Russia favored an understanding or agreement in which the Ukraine could trade with both East and West, Russia and Europe. This arrangement was consistent with the original understanding that Ukraine would be neutral. This plan appeared to be in principle reasonable but was met with opposition in the West in part because it would inevitably mean Russia would supply Europe with large quantities of natural gas and oil and could supplant or undersell western oil producers. Russia might very well have preferred closer ties to the West if the security and trade issues had been resolved.

From the perspective of one vision of U.S. national interests, there are 2 major problems with Russia supplying Europe and others with large quantities of gas and oil. First, Russian production would drive down the oil prices and make domestic production in the U.S. more difficult. U.S. producers cannot effectively produce at a market price below $100/barrel. Russian oil production at $60 a barrel would put economic and price pressure on domestic U.S. production. In one respect

Russian gas and oil would be extremely beneficial to Europe because it could supply them with cheaper energy. The second problem was that gas and oil sales would provide Russia with large amounts of money which it could use to arm and finance groups such as Iran or North Korea that are antithetical to a secure and stable rules based international order. Russia's closer ties to these regimes, not to mention China, are at least in part a response to the actions we have taken against Russia.

In the period of time between 2014 when Russian troops entered the Crimea and parts of the Donbas, the DPR and LPR, and 2022 when they entered other regions in the Ukraine there was an 8-year period of time in which Russia was still open to the Minsk accord being the basis of a long term security agreement. Throughout this period, Russia has consistently advocated for a regional security agreement that would address its concerns about the eastward expansion of NATO. Had we engaged with Russia in good faith who knows what might have happened. When it is clear that our own current regime does not act in good faith in its dealing with the American people why should we believe that they do act in good faith in dealing with other nations?

There are several major issues in the Ukraine that have to be dealt with. One issue is the Ukraine's future. Could Ukrainian neutrality be acceptable to all sides? Is neutrality a trap or a ruse as the Ukrainians and many others claim. The 2nd problem was the status of areas Russia has already controlled since 2014. Up until about December of 2021 or even March of 2022, Russia was prepared for the DPR and LPR to be independent territories not formally incorporated into the Russian Federation and still subject to Ukrainian sovereignty.

In December 2021, Russia proposed a security agreement that would address its concerns about the Eastward of expansion of NATO and other issues. The essence of Russia's proposal, which the Ukraine and the West called an ultimatum, was an agreement in writing that the Ukraine and former republics of the Soviet Union would not join NATO and that there be no further expansion of NATO in eastern Europe. Russia also wanted, as part of an overall security agreement, recognition of Russian sovereignty in the Crimea. The proposal also called for mutual bans on intermediate range missiles in the Ukraine theater. There were other proposals too but those were the most difficult ones for NATO.

The proposals were considered and in January 2022 the U.S. and NATO issued communiques rejecting the proposals on the grounds that NATO and the Ukraine's independence and sovereignty could not be constrained by Russia. In February of 2022 there was a security conference in Munich. Vice President Harris attended the conference and made comments that could easily have been interpreted as supporting or encouraging Ukrainian membership in NATO. The vice President described Ukrainian membership as a form of deterrence and repeated Ukrainian slogans that Russia could not dictate to the Ukraine whether or not it could be in NATO. This was another bit of evidence at a very critical point that the United States was not willing to take Russia's security concerns seriously. There was no way there could be a security agreement that included Ukrainian membership in NATO.

Ukrainian neutrality was good enough for Russia, at least on paper, but it was not good enough for us. After nearly 8 years of trying to reach a peaceful resolution, it was clear the West had no serious, good faith intention of agreeing with the Minsk Accords. Some of the NATO leaders have admitted as much. Moreover, NATO was engaging in a buildup of military infrastructure in the Ukraine including fortifications even though the Ukraine was not a member of NATO. It appeared certainly possible that the West was setting the stage for Ukrainian membership in NATO which we knew would very likely result in the outbreak of a major war. Certainly, powerful elements in the West including Biden's handlers were pushing for Ukrainian membership in NATO.

Three days after the Munich conference, Russia sent regular military forces into other parts of Donbas region and Ukraine for what it called a special military operation. Russia also recognized the DPR and LPR as independent States. Russia also took action against secret NATO installations in the Ukraine that did not officially exist. Russia took out an underground command bunker killing more than 50 NATO military officers.

The Biden regime responded by taking several actions. They seized $360 billion of Russian money that was in the SWIFT system, the U.S. dollar-based system for settling international transactions. They also seized private property of Russian oligarchs in the form of their yachts without any process or lawful authority.

The overthrow of Yanukovych was the Molotov cocktail that preceded the secession of Donetsk and Lugansk from the Ukraine. Had we acted in good faith in not supporting the overthrow of Yanukovych and made a reasonable effort to take Russia's security concerns seriously it is possible there would be no war now or at least a ceasefire could have been negotiated. We could also probably have much better relations with Russia.

In March of 2022 just after the start of the war, there was a peace conference in Istanbul. Russia in essence agreed to the same proposals it had made in December and also further agreed to a plebiscite in Crimea. The people in Crimea almost certainly would (and did) vote to remain in the Russian Federation rather than the Ukraine. The Ukraine would be able to keep its sovereignty and Odessa the Ukraine's Black sea port. At that time, Russia was still willing for the disputed areas the DPR and LPR to be administered by the OCSE. Ukraine was also prepared to agree with this proposal. By some reports, the Ukraine had initialed approval of preliminary documents. This proposal gave the West and the Ukraine a face-saving way to retreat from our extreme stance of insisting that Russia withdraw from disputed areas. This plan would avoid a potentially catastrophic conflict and still allow Ukraine to retain its sovereignty and its Black sea port. Both the British and also Biden insisted this proposal be rejected. In part it may have been felt that to accept the proposal would reward aggression. Ukrainian rejection of this proposal may have been due to threats from the U.S. to cut off further aid if Ukraine accepted the terms of a negotiated agreement. It is very difficult to understand why Zelinsky and the Ukrainians would reject this fairly reasonable proposal without assurances from Biden that the United States and NATO would support the Ukraine militarily including with combat troops to expel Russia from disputed areas of the Ukraine or to at least protect the Ukraine from military reoccupation.

Intransigence combined with weakness is not a good combination in any setting especially international diplomacy. All the parties in the Istanbul negotiations had to have understood a few basic points.

1) Joe Biden did not have the authority to bind the United States to any commitment involving the military security of the Ukraine.

2) The Ukraine has no credible ability on its own to militarily defend itself if Russia decides to occupy the rest of the Ukraine.

3) Current U.S. policy of refusing to support a negotiated settlement including written assurances that the Ukraine will not join NATO leaves Russia with limited options.

These are complicated, serious problems of course. There are issues involving trade, collective security, long term relations with Russia and an increasingly serious risk of a major war breaking out. There may well be validity to the Ukrainian position that neutrality is a trap, that Russia will never truly allow for Ukrainian independence. Many people sympathize with Ukrainian aspirations for complete independence including membership in NATO. However, the situation is not so simple. The situation is much more dangerous now in part because the Russian regime appears to no longer have an interest in closer ties to the West.

In May and June of 2024, we acquiesced in the Ukraine's use of long-range missiles that we had supplied to attack targets inside Russia itself. Russia has warned repeatedly that if Ukrainian missiles are sent into Russia itself as opposed to the disputed areas, Russia may respond by attacking NATO bases outside of the Ukraine. This response would be consistent with international law. Joe Biden's response to these warnings has been to supply the Ukraine with missiles and in late May of 2024 allow the Ukraine to shoot missiles into Russia itself. Biden also authorized the use of American and NATO aircraft to conduct bombing missions inside Russia itself. In June of 2024 there was a missile attack on the Crimea that appears to have been NATO supported and resulted in numerous civilian casualties. This escalation is totally inconsistent with a good faith effort to find a peaceful resolution and is occurring at a time when Russia is still willing to discuss a peaceful resolution.

American policy under Biden from the beginning has never been to actually militarily secure the disputed areas or to achieve a peaceful resolution. We certainly know now, and should have known in the beginning, that expelling Russia from the Russian speaking areas of the Donbas or the Crimea is not an achievable goal. A military victory over Russia on Russian soil is clearly impossible. The policy has been to wear down Russian forces and make them pay a very high price for sending its military into disputed areas of the former Soviet Union in what began as a limited special military operation. Another purpose of American policy was to keep the Russian military preoccupied. The problem with this policy is that it is not consistent with a good faith effort to reach a peaceful settlement. Our policy gives Russia less incentive to negotiate especially if we won't negotiate with them. In fact it gives them an incentive to reoccupy the rest of Ukraine.

In July of 2024 in Russia changed its negotiating position somewhat. Russia was still potentially willing to agree to a resolution that leaves Ukrainian sovereignty intact and allows the Ukraine to retain the port of Odessa. Russia withdrew the offer of another plebiscite in Crimea. They required Ukrainian forces to withdraw from the disputed areas of the Donbas, including Kherson and Zaporizhzhia and to withdraw its application to join NATO. The requirement that the Ukraine withdraw from the disputed territories is reasonable in view of the totality of the circumstances. There will never be a negotiated settlement that involves Russia withdrawing from the Crimea or the disputed territories or at least the DPR and LPR. Any final settlement that leaves the Ukraine intact will require written guarantees that the Ukraine will not join NATO. That is why the Russian demand that the Ukraine withdraw its application to join NATO is also reasonable. That step is necessary to de escalate the situation. However, Russia was not demanding written guarantees as a pre-condition of negotiations. This was a very positive development because it provided a way to avoid a potentially catastrophic conflict. Russia has also so far not demanded release of the $360 billion we seized. Even now, Russia probably does not want to occupy all of the Ukraine let alone Europe.

There is significant internal instability within the Ukraine as evidenced by the serious questions as to who the head of State should be. Zelinsky's term as President of the Ukraine ended in

May 2024. Zelinsky has announced that not only will there not be elections due to the war but also the provisions of the Ukrainian constitution which confer Presidential authority on the leader of the Ukrainian Parliament, Ruslan Stefanchuk, will not be respected. Many countries in the world do not diplomatically recognize the Zelinsky regime.

Significant portions of the arms and military equipment we supply to the Ukraine are not used in an effort to expel Russians from the areas Russia has occupied since 2014. The Ukraine knows that is a futile effort that will never happen. The arms are used to defend against further advances by Russia and also against people living in the Ukraine who are considered collaborators or traitors. In other words, people who would prefer to be in an area associated with the Russian Federation rather than the Ukraine. Western supplied arms have recently been used to invade Russia itself in the Kursk region.

Secretary Blinken in the summer of 2024 confidently announced that Ukraine would join NATO. This development would be a radical escalation of our previous policy of seeking to cause as much damage to Russia as possible without a direct confrontation with Russia. Our leaders certainly understand that military victory for Ukraine without significant NATO support is not possible. The Ukraine joining NATO would virtually guarantee a direct clash between NATO and the Russian Federation and further would provide Russia a very strong incentive to invade the remaining parts of Ukraine that they had heretofore left intact in hopes of a resolution.

By the summer of 2024 it was clear that the Ukraine was militarily and strategically defeated. There is no credible path to military victory that does not include direct involvement of NATO forces and even that will likely further compound the disaster even more. What is victory in Ukraine? Victory in the current state of affairs, realistically, would be Ukraine retains its sovereignty as a neutral country minus Crimea and the DPR and LPR and avoids reoccupation of the Ukraine by Russian Federation forces. Russia is continually taking more and more territory in the Ukraine.

Russia has been extremely restrained in its direct military responses to NATO and the West so far. They haven't responded, yet, to us blowing up the Nordstream pipeline. They haven't responded to several NATO sponsored attacks within Russia. The recent deployment of North Korean troops in the Ukraine theater is a clear signal that this crisis must be peacefully resolved soon. A direct confrontation between NATO and combined Russian and

North Korean forces could very easily become a worldwide conflict. North Korea has already bombed its bridges to South Korea. The Chinese leader recently told his people and military to prepare for imminent war with the United States.

It is important for the American people to recognize that Putin, as imperfect and dangerous as he is, is not a communist. The communist party is an opposition party within the Russian Federation. Putin is under significant internal pressure from Russian nationalists and the communists. If we continue to sponsor attacks within Russia, Putin will be forced to respond or he himself may be deposed and replaced by more extreme elements. At some point Russia will respond if a peace agreement is not reached.

In August of 2024, while there were still peace negotiations going on, the Ukraine with indirect NATO support sent troops and tanks into Russia itself into the Kursk region in an effort to capture a nuclear power plant and use the threat to blow it up as a bargaining chip. Once this offensive is rolled back, the Ukraine on its own will have little viable military defenses left. Russia has already intimated that once the Ukrainian military is defeated there won't be any reason for negotiations. There won't be anything to talk about. The only terms will be unconditional surrender. There is certainly support for this position within Russia.

During the June 2024 debate between Trump and Biden, in discussing the Ukraine situation, Biden spoke almost incoherently and said in effect if the Russians take Kiev, "then we would see a war" and associated this somehow with Article 5 of the NATO charter. As if the Ukraine is part of NATO and the start of World War III was no big deal and certainly nothing that we should negotiate about. Our policy of refusing to discuss this situation in a meaningful way increases the chance that Russia may eventually take Kiev even though that may not have been their original goal or desire.

We had walked away from the negotiating table and although there may have been a chance of reviving negotiations the Ukraine chose instead to launch a horrific terrorist attack against young civilians, the Crocus theater attack which killed well over 100 Russian young people at a concert in the spring of 2024.

In September of 2024, while there was still hope of a peaceful resolution, Biden's controllers announced that the United States would authorize and participate in launching sophisticated cruise missiles into Russia itself. This development caused a major disruption of peace negotiations and resulted in Russia changing its military doctrine as to first use of nuclear weapons. Current Russian policy is that if Russia is attacked by a non-nuclear armed State (i.e. Ukraine) using weapons and military support from a nuclear armed state (i.e. the United States), Russia will regard this as an act of war by a nuclear armed state against Russia. Russia's response in that situation may include the full range of Russia's military capabilities. Russia could easily cause enormous damage to the United States without even using nuclear weapons.

Joe Biden and his handlers do not have the authority under our Constitution to involve us in a direct conflict with Russia without Congressional approval. The provisions of the War Powers act have already been triggered by the Ukraine shooting U.S. supplied missiles into Russia and actually invading Russia itself. Congress should demand that Joe Biden immediately present to Congress his case as to why WWIII is necessary and appropriate and why America should risk its very existence to liberate the Crimea from Russian rule. Biden or Harris if she is elected, should be required to seek an appropriate declaration of war and, if a declaration of war isn't obtained from Congress,

President Biden should stop escalating this crisis and immediately and in real good faith seek a peaceful resolution.

A couple of observations about the situation in the Ukraine are that are undeniable since the spring of 2024 when the Congress voted to borrow and print another $61 billion and give it to the Ukrainian regime.

1) There is no credible path to military victory that does not involve NATO forces and the very near certainty that a much wider war will break out.

2) Of the money we have given so far only about $50 billion has gone to military expenses. The rest is unknown. Congress has resisted efforts to provide an accounting.

3) The American Republic as it is currently constituted has no possibility of surviving WWIII intact.

World War III is unthinkable. It cannot and must not happen. Even though some say WWIII has already started there may still be a way to avoid even more extreme hostilities. The risk is simply too high. Nuclear weapons would almost certainly be used probably by us.

As complicated and as dangerous as the situation in the Ukraine is we cannot risk Americas existence over what is in many respects a civil war within the former Soviet Union. We should get our own house in order. How are we going to defend the Ukraine from Russia when we can't even defend ourselves from the armed criminal gangs taking over apartment buildings and drug cartels that are killing thousands of people here.

What is the way forward at this point to avoid WWIII. First, we have to recognize that our policy so far has been misguided and inconsistent with a good faith effort to achieve a peaceful resolution. We have been using NATO in an aggressive manner. We have meddled in Ukrainian affairs and not respected Ukrainian sovereignty. A large part of the reason why this situation has gotten out of control is that we continually tried to expand NATO eastward contrary to understandings we had reached with Russia and despite their repeated warnings. Russia has been willing to agree with Ukrainian neutrality. We are the ones who have not agreed with Ukrainian neutrality.

We simply have to accept that our 35 year long effort to expand NATO to Russia's doorstep has failed. Not only has it failed but the failure has come at a very high cost. The cost is not just the hundreds of billions of dollars we have pumped into the

Ukraine. The far more costly development has been the strategic realignment of Russia and China leaving the United States currently isolated and very over exposed. Second, the Ukraine has been militarily defeated. That situation is not going to change no matter how much money we print or borrow to pump in there. There may still be a possibility of saving the remaining parts of the Ukraine and avoiding an even more catastrophic conflict. A peace agreement that is reasonable may still be within reach given appropriate responsible leadership in the West particularly the United States.

It may already be too late to avoid a situation in which the West is faced with a choice: Surrender the Ukraine or escalate to WWIII. Note for future reference: This is one of the reasons why it is not a good idea for a President to accept millions of dollars from a belligerent in a conflict that could lead to WWIII.[11] Russia's negotiating position understandably changed from being open to continued existence of Ukrainian sovereignty under negotiated conditions, even after the start of the war, to being one of insisting on what they view as a reasonable security agreement before the war can be concluded. It was pretty clear in the beginning that this impasse would be the end result if Ukraine continued to

---

[11] Biden Family Investigation - United States House Committee on Oversight and Accountability

refuse to negotiate a diplomatic solution. Nevertheless, we must try.

We should communicate immediately with the Russian government the willingness of the United States and NATO to de-escalate this situation and seek a peaceful resolution. In connection with this effort several other steps would be very helpful.

1) We consider recognizing the head of the Ukrainian parliament, the Rada, as the current legitimate head of State of the Ukraine or at least encourage Zelinsky and Stefanchuk to work together to resolve the matter of who should be the head of State and create a timeline for the restoration of Ukrainian sovereignty.

2) We publicly announce that the plebiscites previously held in the disputed Russian speaking areas of Lugansk, Donestk and the Crimea were legitimate, and those people actually do want to disassociate from the Ukraine.

3) We further indicate to Russia that in connection with a final agreement that respects Ukrainian sovereignty including Odessa, the Ukraine and the West will:

4) Recognize Russian sovereignty in the Crimea;

5) We agree with the Minsk II accord framework that provides for Ukrainian neutrality. In other words Ukraine could trade with both East and West and not be part of NATO.

6) We announce that we are willing to put in writing that Ukraine could not join NATO and NATO military forces or weapons would not be stationed in the Ukraine.

7) We agree to withdraw U.S. and NATO forces from the Ukraine including any bio weapons labs and CIA stations.

8) We agree to release the $360 billion we seized through the SWIFT system from Russia but have not yet actually stolen.

9) We and the Ukraine agree to not seek any form of reparations from Russia if they agree to not seek reparations from us for blowing up the Nordstream pipeline.

It is time for us to listen and start acting in good faith while there is still a possibility of avoiding a catastrophic war. Although we can never get back to the terms we walked away from, we still may be able to avoid WWIII and also preserve Ukrainian sovereignty. We are not going to be able to get everything we want. There is no inherent reason why we cannot accommodate Ukrainian neutrality and free trade with both East and West. Yes, there are some problems with that and yes the cost now is much higher than it would have been in 2021. Our current negotiating position may leave the Russians little choice but to eliminate Ukrainian sovereignty. That is the only way they can guarantee that Ukraine does not put more NATO forces on Russia's borders and continue to launch missiles, tanks, troops and jets into Russia. Our official position until fairly recently was that we can't negotiate until Russia withdraws from the disputed territories. This position cannot be described as being a reasonable good faith effort to contribute to a peaceful resolution of this very dangerous and escalating situation.

CONCLUSION

The American Republic is on life support. 2024 is probably the last opportunity for the American people to re assert control of the American Republic; restore legitimate good faith Constitutional Government and restrain the globalists and math deniers that have hijacked the dollar. The clues are everywhere the that the American Republic is facing an existential internal threat. In looking at the totality of the circumstances, the enormity of what the deep state enablers have done to America; to the dollar; to our security; to our sovereignty; to facilitating an invasion of the United States; to the weaponization of our justice system; to the surrender of massive amounts of military equipment; to risking World War III; to purging the military; to the massive and concerted efforts to destroy anyone who dares to fight back in a meaningful way and finally to at least 2 attempted assassinations of President Trump, the conclusion is inescapable that the American Republic is under attack and has already been dealt a serious blow.

It is almost inconceivable that good faith incompetence is a credible explanation of the Biden administrations and deep state actions. An average American citizen acting in good faith could do a far better job than Kamala Harris has at defending our border. Yes, Joe Biden is medically and morally unfit, incompetent and corrupt but the problem is much deeper than that. The actions of Bidens controllers are deliberate and willful not reckless and incompetent. The reason this issue is important is that it is relevant to the scope and urgency of the response of the American people. Biden and his parties actions are designed end America's existence as a free sovereign Republic and transform America from a Constitutional Republic based on citizen sovereignty to a post Constitutional, one party state based ultimately on non-citizen sovereignty.

As painful and as damaging to the United States as the Biden/Harris deep state regime has been, perhaps it has been a blessing in disguise, at least if WWIII does not start, because it has revealed the depths of the corruption that exist at the national level and the stranglehold that the deep state and globalists have on the American people and the American dollar. The obvious bad faith and 2 tier system of justice is out in the open and no longer hidden from the American people. So much of the corruption was hard for the average person to even comprehend. For years the media has not provided the American people with reliable information to put events into an understandable context. It never occurred to the average American until the last few years that the media would refuse to report events honestly to the American People. It never occurred to the average American that many elements within the Republican party could be so corrupt that they would put their own personal desire for money and power ahead of the National interest. Who could have imagined as recently as 4 years ago that so much of our establishment could be cowed into accepting an uncertifiable and fraudulent national election of critical importance. Who could have imagined that our criminal justice and legal system could be perverted and used to impeach a President and attack, prosecute and even imprison a candidate that could not be beaten in a free and fair election. Who could have imagined that America would again suffer an assassination attempt on the President.

Perhaps November 2020 was a new episode in a story that started in November of 1963 and not so much a coup that was totally unprecedented. Maybe it's a new story altogether. We did have a *coup d'etat* in 2020. The evidence is plain as day, right in front of our face. It can no longer be reasonably denied. In the fog of this current crisis it will not be possible to sort all the details before the 2024 election.

After the June 2024 debate when it became obvious that Biden is totally unfit for office the Democrat party continued to lie to the American people while they frantically searched for a way to dump Biden. Top party leadership made statements to the effect that Joe had a bad day but they were backing him. Not a peep about the extreme danger to the American people and frankly the world that is presented by having a dementia patient sitting in the oval office. All of them told the American people that there is no problem with Joe staying in office not only until January 2025 but beyond that as well. Not a peep about how this fraud against the American people was covered up. They were telling the American people, "don't believe your eyes, don't worry about the extreme risk, put party ahead of country and national security and put us back in power again". If the party leadership isn't willing to be honest and act in good faith about something as obvious as Biden's fitness for office what are they willing to be honest about? The answer is virtually nothing. Their entire program either domestically or internationally does not involve honesty or good faith.

Democrat party hacks, the prestituite corps and media propagandists constantly harp about the "threat to our democracy". When they talk about "our democracy" they are not talking about the United States as a sovereign Constitutional

Republic. They are talking about the deep state, this little thing of ours, the regime of all the people and groups throughout the world who want to spend Americas wealth and deploy Americas military. America is not a democracy. America never has been and never will be a democracy. America is a Constitutional Republic. We will either retain our status as a sovereign Constitutional Republic or we will lose it an enter a post Constitutional era that may temporarily have some quasi democratic features before eventually becoming a dictatorship like Venesualza or Cuba but it will not be a sovereign Constitutional Republic any longer.

We may be entering a post Constitutional era. Actions by both the Democrat party leaders and powerful elements within the Republican party have breached our social compact by both legal and extra legal actions. Moreover, many politicians are refusing to act in good faith. Their actions are designed to further their political agendas not to preserve, protect and defend the Constitution of the United States.

There is only one Donald Trump. His supporters are not a personality cult. They are sincere, decent American people acting in good faith and concerned about the extreme danger our Republic is currently facing. He is not perfect. He is not right about everything. He is the person at this moment of history uniquely qualified to serve the American people in the sacred duty as the true President.

If the American people stand idly by and allow a patriot of the stature of President Donald Trump to be abused, attacked and imprisoned there is little hope that our Republic can survive intact. We will not be a free sovereign people anymore. We will become a bunch of former citizens and intimidated serfs waiting for Americas post Constitutional era to unambiguously begin.

Nothing said herein should be construed as a call to violence. Our Republic can still be saved through nonviolent and Constitutional means. We cannot depend only on Donald Trump. It is our responsibility as citizens to save our Republic or we will lose it. The Republican party, despite its flaws, is the only major party that retains the moral high ground and by its actions remains faithful to the Constitution. It is critical to the survival of the American Republic that President Trump be re-elected. And, almost as critical, the Democrat/deep state party must not be allowed to control either the House or the Senate. The deep state makes no secret of its desire to use any form of lawfare it can including initiating a Constitutional crisis, to deny the American people their Constitutional right to choose their President. As soon as they can, if they have power, the Democrats will attack the Supreme Court.

All patriots, independents and any citizen not committed to the destruction of the American Republic must register and vote not only for President Trump but also for down ballot Republican candidates for the House and Senate. The millions of Christians and gun owners who have not voted in large numbers in past elections must register and vote. The patriots all over the county and especially in the battle ground States and Texas must find a way to cast their ballot. The turnout of the patriot electorate must be massive so that the result is "too big to rig".

There is no easy way anymore. The people and forces attacking President Trump and his supporters in bad faith using lawfare and abusing the public trust are the enemies of our Republic. The enablers may be acting wittingly or unwittingly. We cannot make them act in good faith only they can do that. We can, however, peacefully prevent them from attaining political power. That we must do. They have to be stopped or they will destroy our Republic. The problem doesn't get any more complicated than that.

In his acceptance speech for the Republican nomination for the U.S. Senate in Minnesota, Royce White stated this concept very succinctly when he said the issue in November of 2024 facing the American people is a simple one: Are we going to continue to have a viable country or are we not. The issue really is that simple. We can have peace, prosperity, security, freedom and liberty or we can have the current Democrat party in power. We can no longer have both. The time to choose has come.

BIBLIOGRAPHY

1. Allen, Joe <u>Dark Aeon: Transhumanism and the War Against Humanity</u>, (War Room Books, 2023).

2. Conquest, Robert <u>Harvest of Sorrows</u>, (Oxford University Press 1987).

3. Hoft, Joe <u>The Steal Volume 1: Setting the Stage</u>, (2022).

4. Hemingway, Mollie <u>Rigged: How the Media, Big Tech and the Democrats Seized our Elections</u>, (Regnery 2022).

5. Sontag, Raymond James and Beddie, James ed. <u>Nazi-Soviet Relations 1939-1941</u>, (Didier, 1948).

6. <u>Summary of Election Fraud in the 2020 Presidential Election in the Swing States</u>, (Published Online) and sources cited therein.

7. Wallach, Jehuda L. <u>The Dogma of the Battle of Annihilation</u>, (Greenwood Press, 1986).

8. Weinburg, Gerhard L. <u>The Foreign Policy of Hitlers' Germany</u>, (University of Chicago Press, 1980).

ABOUT THE AUTHOR

Benjamin Casad, B.A. University of Kansas 1989, J.D. University of Kansas 1992, studied modern European history, law and international law. Other areas of study include accounting and mediation. The author is an attorney and longtime observer of current affairs. This is the authors first published book.

www.ingramcontent.com/pod-product-compliance
Lightning Source LLC
Chambersburg PA
CBHW052114030426
42335CB00025B/2973